Alberto Caldas-Alvarez

Atmospheric Moisture Effects on Deep Convection in the Western Mediterranean

Wissenschaftliche Berichte des Instituts für Meteorologie und Klimaforschung des Karlsruher Instituts für Technologie (KIT)
Band 81

Herausgeber: Prof. Dr. Ch. Kottmeier

Institut für Meteorologie und Klimaforschung
am Karlsruher Institut für Technologie (KIT)
Kaiserstr. 12, 76128 Karlsruhe

Eine Übersicht aller bisher in dieser Schriftenreihe
erschienenen Bände finden Sie am Ende des Buches.

Atmospheric Moisture Effects on Deep Convection in the Western Mediterranean

by

Alberto Caldas-Alvarez

KIT Scientific Publishing

Karlsruher Institut für Technologie
Institut für Meteorologie und Klimaforschung

Atmospheric Moisture Effects on
Deep Convection in the Western Mediterranean

Zur Erlangung des akademischen Grades eines Doktors der Naturwissenschaften von der KIT-Fakultät für Physik des Karlsruher Instituts für Technologie (KIT) genehmigte Dissertation

von Dipl.-Phys. Alberto Caldas-Alvarez

Tag der mündlichen Prüfung: 12. Februar 2019
Referent: Prof. Dr. Christoph Kottmeier
Korreferent: Prof. Dr. Peter Knippertz

Impressum

KIT Scientific Publishing

Karlsruher Institut für Technologie (KIT)
KIT Scientific Publishing
Straße am Forum 2
D-76131 Karlsruhe

KIT Scientific Publishing is a registered trademark
of Karlsruhe Institute of Technology.
Reprint using the book cover is not allowed.

www.ksp.kit.edu

Print on Demand 2019 – Gedruckt auf FSC-zertifiziertem Papier

ISSN 0179-5619
ISBN 978-3-7315-0947-9
DOI 10.5445/KSP/1000097100

Abstract

The presented investigation focuses on the role of atmospheric water vapour in shaping the intensity, structure, location and timing of atmospheric convection. The aim of this thesis is, therefore, gaining knowledge and reducing uncertainty associated with the interaction between atmospheric moisture and convection, assessing the impact of nudging Global Positioning System (GPS)-derived Zenith Total Delays (ZTDs) in different resolution simulations of heavy precipitation from seasonal to event scales.

Analysis of atmospheric simulations in a realistic set-up with the well-established Consortium for Small-scale Modelling (COSMO) model are carried out on different temporal and spatial scales. Simulations from seasonal to event scale are produced to improve the understanding of the mechanisms of convection and moisture interaction. A unique GPS-derived ZTD data set, a proxy of column atmospheric moisture, merging data from more than 25 European networks, with a temporal resolution of ten minutes is nudged to improve the representation of the atmospheric humidity distribution. The moisture nudging sensitivity experiments are done on different horizontal resolution simulations (7 km, 2.8 km and 500 m) to assess the sensitivity of the model in different configurations, from convection parameterized to convection-permitting, reaching the micro-α scale.

Nudging column atmospheric moisture improved the humidity wet bias present in COSMO in the simulation of the autumn 2012 period in the WMed. The root mean square error, the agreement index and the mean absolute error scores improved in seasonal and event scale simulations using 7 km, 2.8 km and 500 m grid spacings. Past studies have also reported a wet bias in COSMO (Schraff et al., 2008; Cress et al., 2012; Devidasrao, 2012). The column moisture nudging could not, however, correct biases over particular high-terrain stations (e.g. the Alps, the Pyrenees) given the misrepresentation of the model surface height. Neither could the column moisture nudging correct humidity biases which remained in the order of 1 $g\ kg^{-1}$ at particular pressure levels. Errors in this order in the vertical distribution have already been reported as crucial for convective precipitation representation (Weckwerth et al., 1996; Khodayar et al., 2016b; Chazette et al., 2015b). This issue, present in all grid length types, could be overcome by jointly nudging column moisture observations with operational radiosondes. This was demonstrated for a heavy precipitation event of the autumn 2012 period.

The autumn 2012 seasonal mean values, endured an Integrated Water Vapour (IWV) reduction between -3% and -8% as a result of the column moisture nudging. The largest seasonal IWV reduction took place over the Iberian Peninsula and Italy (IT) and the lowest over France (FR) .The outcome for instability as given by CAPE and for the total cloud cover was an intense reduction of about $-10\ J\ kg^{-1}$ and of -4%, respectively. All in all, these impacts reduced the total precipitation amount over land (with decreases ranging between -9% and -65% which were largest for FR and shortest for IT. Regarding the dependence on the used model grid-spacing, the precipitation reduction was larger in the 7 km simulations (between -23% and -65%) than in the 2.8 km simulations (between -9% and -34%).

The analysis on the seasonal conditions as well as of two Heavy Precipitation Events (HPEs) of the period showed that the convective processes enduring the largest impact to the column moisture nudging were instability and drying (moistening) of the lower free-troposphere. The latter had stringent impacts on entrainment/detrainment processes and buoyancy reduction (intensification) of the convective plume. Besides, to a lower extent, the humidity corrections affected the distribution of the large-scale pressure distribution with a tendency for higher geopotential heights at 500 hPa and surface pressure. This effect was observed over France only, due to the larger number of GPS stations and that typically low pressure systems are located over France during autumn.

About 90% of the HPEs in the autumn 2012 period were found to be preceded by IWV monotonic increases. In the mean, these increases were of 4 mm (with a standard deviation of 5mm) and lasted for 9 h in mean (9 h of standard deviation). When the most extreme precipitation events were considered (above percentile-90), the IWV increases preceding precipitation were largest (>10mm) and occurred for longer time periods ($\approx$ 30 h). These results apply to the simulations using a 7 km and a 2.8 km grid over local target areas O(10^5 m) in the north WMed basin.

The impact of simulating errors in the vertical distribution of humidity was studied in detail, simulating an HPE taking place on the 24-Sep-2012 over southern France jointly nudging observations of column atmospheric moisture and operational radiosondes. Model grid-spacings of 7 km, 2.8 km and 500 m were employed in this experiment. The results show that the humidity representation in the vertical distribution, as well as IWV are improved when applying the nudging of column moisture and radiosondes in a 500m grid. Especially for specific humidity below 900hPa. Moreover, nudging radiosonde information improves the representation of the structure, location and amount of precipitation over the affected areas, for every used grid-spacing type.

This thesis provides novel insights on how IWV increases precondition heavy precipitation, the humidity wet bias in COSMO in the period, the atmospheric moisture impacts in convective processes and precipitation evolution and the improvement in moisture and precipitation representation by nudging GPS data and radiosondes in 500 m simulations. It provides deeper knowledge on COSMO responses and biases useful for users and developers of numerical prediction systems. The presented results demonstrate the benefit of using column moisture observations to correct the humidity biases with a large impact on precipitation.

Contents

1 Introduction and Relevance of the Thesis

Heavy precipitation, is a weather phenomenon with a high impact in communities and populations worldwide. In the Western Mediterranean (WMed) region, heavy precipitation is particularly damaging and frequent during the autumn season in the WMed. The European Environment Agency (EEA, 2016) estimated that more than 3000 people died in the WMed and northern Europe, considering only storm events, in the period 1980-2013. In addition, this high impact weather entailed monetary losses up to 148.000 million EUR. When the very intense downpours take place over saturated sub-soils, areas of large ground-water content or poorly urbanized areas, heavy precipitation often evolves to flooding, at times, in the course of just a few hours (flash flooding). Specially affected by these events are WMed countries such as Algeria, Italy, France and Spain as shown by the analysis of the last years trends

Figure 1.1: Flooding in Sant Llorenç des Cardassar, Mallorca, Spain (9-Oct-2018). Maximum precipitation rate of 90 $mm\ h^{-1}$. Caldera, Cati, 2018, EFE, *https://elpais.com*, last accessed 08-Nov-2018.

(Gaume et al., 2016). Reducing the number of casualties and losses is a priority for the Mediterranean and European national weather services, responsible for issuing the corresponding warnings and alerts to the emergency units and population. Hence, Numerical Weather Prediction (NWP) systems applied to nowcasting and medium range forecasting are fundamental tools in the prevention and mitigation of the impacts of extreme precipitation, which benefit from an improved understanding of the underlying mechanisms. In spite of the advancements made in the last decades in reducing the uncertainty of forecasting such events (WMO, 2012), their representation remains a challenge. This is so, given most precipitating systems bringing these impacts are of convective nature (Ferretti et al., 2014) and may happen very locally, at spatial scales of a few kilometers and swiftly, with precipitation totals of 150 *mm* in the course of a few hours (Davolio et al., 2016; Ducrocq et al., 2016).

Past trends and climate projections show an escalating number of Heavy Precipitation Events (HPEs) and their intensity in warming climates (Sala et al., 2000; Alfieri et al., 2018; IPCC, 2018). Future projections for the WMed predict warming, an overall decrease in precipitation and an increase of the number of extreme precipitation events (Vautard et al., 2014; Greve and Seneviratne, 2015; Drobinski et al., 2016; IPCC, 2018). Lionello and Scarascia (2018) analysed the recent and future evolution of the Mediterranean basin and assessed warming at the basin to be 20% larger than the global average. Tramblay and Somot (2018), using downscaled Regional Climate Model (RCM) simulations from the Euro-Coordinated Regional Climate Downscaling Experiment (CORDEX), found significant trends of extreme precipitation increases (up to +20%) by the end of the twenty-first century in the European coasts of the Mediterranean basin. Two identified factors bringing these changes in extreme precipitation over the Mediterranean are the poleward shift of the circulation and a thermodynamic effect leading to an increase of Integrated Water vapour (IWV) (Drobinski et al., 2016; Pfahl et al., 2017; Tramblay and Somot, 2018). Thermodynamic mechanisms induce vertical transports of mass which bring water vapour toward saturation forming Deep Moist Convection (DMC). DMC contributes to about $70\% - 80\%$ of precipitation in the tropics (Mohr et al., 1999) and it responsible for extreme precipitation events at mid-latitudes. Its study is of growing interest to the scientific community (Coppola et al., 2018), still its understanding remains amorphous (Stevens, 2005; Sherwood et al., 2010; Ahrens and Samson, 2010), this is due to the manifold interactions of the related processes, acting at a broad range of scales, from turbulence to large-scale advections. In this regard improving the understanding of the pathways of the atmospheric moisture and convection interaction is crucial. Previous studies coincide on the need of three ingredients for the befallen of DMC, a conditionally unstable environmental lapse-rate, sufficient moisture and some mesoscale mechanism by which the air parcels are lifted to the Level of Free Convection (LFC) (Doswell et al., 1996). In all of them, water vapour exerts an influence. The spatio-temporal distribution of water vapour affects instability and buoyancy of the lifted air masses, as well as the environmental lapse-rates setting how deep the updraughts will reach (Stevens, 2005; Khodayar, 2009). Initiation of convection depends to a large extent on the low-level moisture distribution and its presence above the Planetary Boundary Layer (PBL) regulates the intensity of convection once it has been triggered (Markowski and Richardson, 2010; Sherwood et al., 2010).

Simulating these mechanisms accurately is still a challenge. Deficiencies in the representation of past events (Ducrocq et al., 2016) as well as discrepancies among different well-established models (Khodayar et al., 2016b; Coppola et al., 2018), hamper the potential for weather hazard prevention and warning issuing. In particular, the misrepresentation of humidity in the models has been pointed out as a source of error. Sherwood et al. (2010), conveyed that the analysed global models dry out excessively the troposphere, whereas Convection Permitting Models (CMPs) have a tendency for an excess of moisture in the boundary layer. Other studies report deficiencies of the order of 2 $g\ kg^{-1}$ (see Fig. 1.2) in specific humidity (Chazette et al., 2015b) between model and observations or large biases in the representation of IWV, e.g. Consortium for Small-scale Modelling (COSMO) and High Resolution Limited Area Model

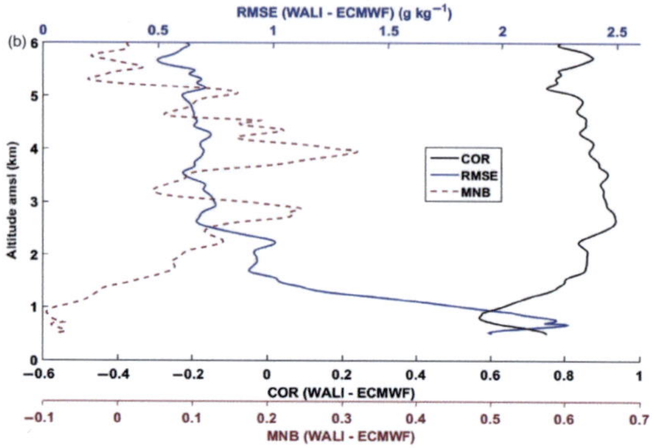

Figure 1.2: Obtained from Chazette et al. (2015b). Vertical profiles of Root Mean Square Error (RMSE) in $g\ kg^{-1}$ (blue), Correlation (COR; black) and Mean Normalized Bias (MNB; sepia) of water vapour mixing ratio retrieval between the ground-based Water-vapour Raman Lidar (WALI) and European Centre for Medium-range Weather Forecasts (ECMWF) reanalyses. The calculations have been made on 51 cloud-free coincidences for the period 17-Sep to 30-Oct, 2012.

(HIRLAM) (Guerova et al., 2016). Errors of this order in the representation of humidity are non-negligible and, as demonstrated by Weckwerth et al. (1996), changes of 1.5 $g\ kg^{-1}$ to 2.5 $g\ kg^{-1}$ exist between the updraft region of a convective roll (showing thermodynamic conditions comparable to those of the cloud base), and the neutral region. Introducing water vapour differences of this order entails large impacts in the representation of HPEs, as shown by past sensitivity studies (Crook, 1996; Keil et al., 2008; Li and Shen, 2010). For example Honda and Kawano (2015) were able to suppress convection in the simulation of a supercell event by reducing 10% relative humidity between 1.7 km and 4.2 km.

In order to overcome errors related to the misrepresentation of atmospheric humidity, national weather services and other institutions are putting efforts in advancing the Data Assimilation (DA) of highly-sampled humidity observations (Guerova et al., 2016). DA is a climate and weather numerical prediction technique by which the fields of the simulated variables are corrected by forcing them toward observed values. An example DA cycle consists of simulations of a past time-window, for instance three hours, where the model fields are corrected towards measurements, followed by a forecast of a future time-window. The overlying idea is that the forecasts depart from a model state which is closest to "reality". Duruisseau et al. (2017), found a positive impact of assimilating 25 water vapour Infrared Sounder (IRS) channels in the 18h lead-time forecast of precipitation. A DA technique implemented in well-established NWP models like COSMO, Fifth-Generation Penn State/BCAR Mesoscale Model (MM5), Bologna Limited-Area Model (BOLAM), MOdello LOCale (MOLOCH), is the *Nudging Scheme*. This technique adds forcing terms, dependent on the differences between the model and the assimilated observations at selected time steps of the run (Schraff and Hess, 2012). The main advantages of the Nudging methodology are its simplicity and its ability to be used with very frequent observations (Guerova et al., 2016). Comparisons against other schemes, for instance, the Kalman filter for convective-scale Data Assimilation (KENDA) system implemented for COSMO, have shown better skill of the Nudging scheme in correcting the humidity biases as compared to KENDA (Schraff et al., 2016).

High-resolution observations of atmospheric water vapour are essential to improve the under-sampling of humidity. This issue can be improved using Global Positioning System (GPS) technology that provides estimates of the Integrated Water vapour (IWV) of the atmospheric column. The retrieval of IWV is made possible due to the delay endured by the emitted L-band microwave signals from GPS constellations. This quantity is called the Zenith Total Delay (ZTD) which is linearly related to IWV. Thus both quantities are interchangeable (Bevis et al., 1992). The use of GPS-derived ZTD or IWV measurements can help overcome the sparse coverage and low temporal resolution of traditional humidity observing systems e.g. radiosondes. Its main advantages are that it is affordable, accurate, it provides very fast observations in the order of minutes and regions such as European countries have a dense coverage (Hdidou et al., 2016; Bock et al., 2016).

In line with the aim of improving the fine-scale representation of moisture variations in the atmosphere, the convective processes and precipitation is dynamical downscaling. There is growing interest by research institutions and operational services in improving the climate and weather simulations at very fine horizontal model resolutions. In this regard a breakthrough in the field has been accomplished by reaching horizontal grid spacings that explicitly resolve deep convection, i.e. grid spacings narrower than ca. 3km, also for long term simulations. Refining the model grid has proven beneficial for the representation of the interaction of the model orography and DMC (Lim et al., 2010; Chan et al., 2012; Prein et al., 2015; Coppola et al., 2018), the representation of the soil-atmosphere feedbacks (Argüeso et al., 2013) or of the convection related indices (Fosser et al., 2016). Reported improvement in the representation of convective precipitation is found for summertime convection, the diurnal cycle, sub-daily precipitation and HPEs (Lim et al., 2010; Kendon et al., 2012; Prein et al., 2015). However, the question is still open as to how fine the grid lengths should be in order to attain the best results for precipitation, (Chan et al., 2012; Barthlott and Hoose, 2015).

Given the relevance of the introduced problem, the aim of this thesis is to gain knowledge and reduce model uncertainty associated with the interaction of atmospheric moisture and convection, assessing the impact of nudging observations of column atmospheric moisture in different resolution simulations of heavy precipitation from seasonal to event scales.

The overall methodology consists on first, assessing the impact on the represented humidity fields, second the impacts on the represented precipitation to end up with the assessment on the changes of the convective variables. For each of these three blocks scientific questions are set out. For instance, *Did the nudging of the column atmospheric moisture bring an improvement in the representation of humidity and precipitation at the seasonal and event scales?*, *How much and for how long does the total column moisture vary over a target area prior to heavy precipitation?* or likewise *What are the different impacts of nudging the model runs toward column moisture observations when using downscaled model horizontal resolutions?*

To answer such questions, modelling experiments on the sensitivity of the well-established COSMO model to corrections of the atmospheric humidity fields have been performed in a realistic set-up. The atmospheric humidity corrections are implemented by nudging every ten minutes a state-of-art GPS-ZTD data set, specially homogenized for the WMed countries, with a dense coverage. An across-scales approach is utilized by seasonal and event-scale simulations performed for the period September to November 2012, in the WMed employing three different horizontal grid lengths (7 km, 2.8 km and 500). These are, to the author's knowledge the first sensitivity experiments focusing on the role of water vapour on DMC, forcing a realistic distribution of humidity using observations of such high temporal resolution and large coverage for a complete season. In addition to the investigation of the whole autumn season of 2012, selected case studies are assessed in detail to further understand the role of atmospheric moisture in the development of convection and heavy precipitation. From these investigations novel insights and quantifications are obtained, on the role and amount of IWV preceding heavy precipitation events, impacts on stability and moisture flux induced by the humidity changes and on the relevance of representing realistically the vertical distribution of humidity.

Chapter 2 introduces the theoretical framework of water vapour transports and atmospheric convection. Chapter 3 summarizes past findings and literature on the characteristics of the WMed region that make it prone to heavy precipitation in autumn. The COSMO model and the Nudging scheme are described in Chapter 4. The set of simulations used, the observational data sets and methods are explained in Chapter 5. Chapter 6 presents the results of the humidity sensitivity experiments regarding IWV and specific humidity for the September to November period on the 7 km and 2.8 km simulations. Chapter 7 gathers the main results of the influence of the water vapour changes in convective phenomena and precipitation likewise for the seasonal simulations. Chapter 8, presents the findings on the IOP6 case study using very high resolution simulations (500 m) and the combined use of ZTD data and operational soundings. Finally, the conclusions are presented in Chapter 9.

2 Theoretical and Phenomenological Background

The theoretical framework of this thesis is introduced in this section. It comprises a summary of the mechanisms of water vapour transport across scales, the main equations, variables and processes of atmospheric convection and a short review of the existing literature on the effects of moisture on convection.

2.1 Atmospheric Water Vapour Across Scales

Water vapour lies at the heart of all key terrestrial atmospheric processes (Sherwood et al., 2010). Its transport takes place across all scales, from global to turbulent (micro-γ). At the global scale, the circulation generally follows a tropics to poles direction and the contribution of tropical moisture accounts for roughly half of the total latent heat at the subtropics and the mid-latitudes (Sherwood et al., 2010). The tropics to poles circulation stems from the large evaporation over the tropical oceans and trade wind regions. The moisture evaporated from land and water parts is advected toward the Intertropical Convergence Zone (ITCZ) where it is transported upward via convection. At the upper levels detrainment of condensed air and clouds takes place due to large-scale circulations transporting the moisture towards higher latitudes. At the subtropics, the air masses reach subsidence regions, where they will descend gradually compressing and warming as they mix with local air.

Besides the global transport of water vapour from the tropics to the poles, moisture also endures transports at the mesoscale. The mesoscale transports are crucial in the western Mediterranean. The moisture distribution at the Mediterranean Basin is largely influenced by the general Ferrel circulation cell, which induces a southerly flow, deflected by the Coriolis force stemming from the Earth's rotation. For this reason, mesoscale moisture transports in the Mediterranean typically have a southwesterly direction. Low and high pressure systems induce cyclonic and anti-cyclonic circulations respectively, which explain the departures from the general southwesterly flows. The water vapour transported to the Mediterranean Basin, associated with heavy precipitation originates mostly at the Atlantic Ocean, the Mediterranean Sea and northern Africa. A review of the state-of-art literature concerning moisture origins for heavy precipitation in the Mediterranean can be found in Section 3.1. At times, the transported moisture may originate at the tropics in a phenomenon of poleward circulation named *atmospheric rivers*. Several studies (Zhu and Newell, 1998; Krichak et al., 2014; Chazette et al., 2015a; Ramos et al., 2018) have reported south to north narrow bands of very moist air, possibly correlated to heavy precipitation over the Mediterranean and the European continent. However, the question is still open as to state whether the moisture really originates at the tropics or if the narrow moist filaments are formed by a moisture sweep up exerted by the cyclone's cold front trying to catch up with the warm front (Dacre et al., 2015).

In order to quantify the water vapour transports over a region, as well as its sinks and sources at the large and medium scales the water vapour balance equation (Eq. 2.1) is used (Rasmusson, 1968; Lamb et al., 2012; González-Rojí et al., 2018). Evapotranspiration is a positive contributor to the atmospheric system, whereas precipitation is the major sink. Moisture transports can either positive or negative contributors. Applied to the large-scale, the contribution of condensation/evaporation due to cloud formation, within the box, is typically neglected since it stands for a low percentage of the total water vapour flux (Rasmusson, 1968). Figure 2.1, illustrates the calculation of the moisture balance at a closed box as in Lamb et al. (2012). The local derivative of the total water vapour content within the box ($\partial(IWV)/\partial t$) depends on precipi-

$$\frac{\partial (IWV)}{\partial t} + \nabla \cdot \boldsymbol{Q} = E - P \tag{2.1a}$$

$$\frac{\partial (IWV)}{\partial t} + \frac{OF}{A} - \frac{IF}{A} = E - P \tag{2.1b}$$

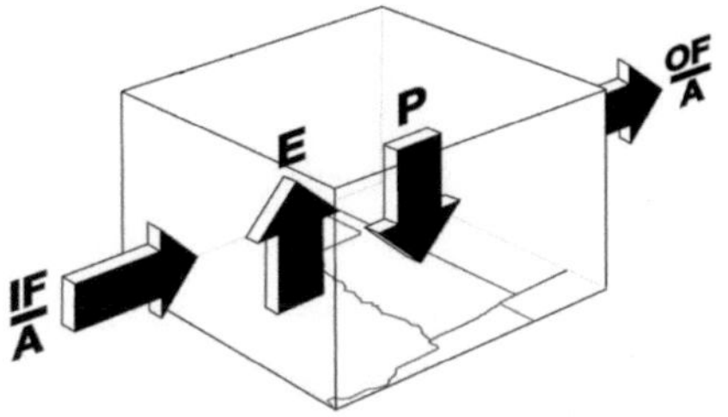

Figure 2.1: Depiction of the contributions to the moisture balance over a closed region by Evaporation (E), Precipitation (P) and total water vapour mass inflow (IF) and outflow (OF) per unit area. Moisture balance equations (left) where the moisture divergence term is split into the inflow and outflow moisture. Obtained from Lamb et al. (2012).

tation (P), evaporation (E) and the moisture divergence ($\nabla \cdot \boldsymbol{Q}$). Lamb et al. (2012) split the moisture divergence term into the total outflow (OF) and the total inflow (IF) per area (see Fig. 2.1).

The moisture advected into the box of study is represented by means of the moisture divergence term (Eq. 2.2aa), which can be split into a thermodynamic and a dynamic contribution to the total divergence (Lamb et al., 2012). Another quantity useful for moisture transport quantification is the moisture flux (Ricard et al., 2012), see Eq. 2.2bb.

$$\nabla \cdot \boldsymbol{Q} = \frac{1}{g} \int_{p_0}^{p1} \boldsymbol{V} \cdot \nabla q \, dp + \frac{1}{g} \int_{p_0}^{p1} q \nabla \cdot \boldsymbol{V} \, dp \tag{2.2a}$$

$$Q_{flux} = -\frac{1}{g} \int_{p0}^{p1} q \boldsymbol{V} dp \tag{2.2b}$$

Where $\boldsymbol{V}$ stands for the horizontal wind vector, q is the specific humidity and p and p_0 are the bottom and top pressure levels for vertical integration and g is the gravity acceleration which is approximated by $\boldsymbol{g} = -g_0 \left(r_e / (r_e + h) \right)^2 \boldsymbol{u}_z$,

with g_0, the standard gravity constant, r_e the Earth's mean radius and h the height above sea level.
The moisture divergence and the moisture flux equations, show the link of the moisture budget and winds. When smaller scales are considered, turbulence processes play an important role in setting the moisture distribution (as it is the case of other scalar quantities such as heat or pollutants, for example). Turbulence can be visualized as a superposition of irregular swirls of different sizes and energy called *eddies* which can be as large as the depth of the boundary layer in the order of $O(10^3)$ m and down to $O(10^1)$ m (Stull, 1988). Ideally, knowing the prognostic equations for moisture turbulent transport would lead to the ability to forecast each of those eddies (Stull, 1988). The turbulent contributions to the flow are known as turbulent departures from the mean and are obtained from the moisture conservation equation. By applying the Reynolds decomposition into mean and turbulent parts, averaging the scalar quantity for mean flow conditions and subtracting it to the derived prognostic equation, the prognostic turbulent equation for moisture has the form

$$\frac{\partial q'}{\partial t} + \overline{U_j}\frac{\partial q'}{\partial x_j} + u'_j\frac{\partial \overline{q}}{\partial x_j} + u'_j\frac{\partial q'}{\partial x_j} = \nu_c\frac{\partial^2 q'}{\partial x_j^2} + \frac{\partial\left(\overline{u'_j q'}\right)}{\partial x_j} \tag{2.3}$$

Where $\partial q'/\partial t$ is the rate of change of turbulent moisture, $\overline{U_j}(\partial q'/\partial x_j)$, is the advection of turbulent humidity by the mean wind, $u'_j(\partial \overline{q}/\partial x_j)$ and $u'_j(\partial q'/\partial x_j)$ are the turbulent advections of mean and turbulent specific humidity, respectively, $\nu_c(\partial^2 q'/\partial x_j^2)$, is the viscous stress term in the transport of turbulent specific humidity, where ν_c is the kinematic viscosity, which has a standard sea-level value of $\nu = 1.46 \cdot 10^{-5}$ m^2 s^{-1} and $\partial \overline{u'_j q'}/\partial x_j$ is the turbulent transport of specific humidity (it describes how q is moved around by the turbulent eddies). Turbulent transports are relevant for example in deepening the moist layer aloft the Planetary Boundary Layer (PBL) which will invigorate convection by preventing excessive dry air entrainment.

Spatio-temporal variability of atmospheric moisture

As shown in the previous section, the atmospheric water vapour transports occur at the large, middle and turbulent scales, and therefore water vapour is one of the most variable components of the atmosphere. Hence, significant improvement in the representation of convection is expected if the humidity fields are simulated realistically with sub-hourly and kilometer accuracy. Steinke (2017) quantified the variability of IWV at small scales to be of 0.6 *mm* under horizontal displacements of 8 to 9 km or time shifts of 15 to 30 minutes over Germany. Variations in IWV of this order are non-negligible, and as will be presented in Section 2.3, such humidity variations impose a large impact in the representation of precipitation amount (Keil et al., 2008; Honda and Kawano, 2015; Lee et al., 2018). Khodayar et al. (2018), assessed in the study of a Heavy Precipitation Event (HPE) occurring during the autumn 2012 season, a variation of 4 $g\ kg^{-1}$ in moisture at the PBL under a 100km horizontal displacement. Similar values of humidity variations were reported by Girolamo et al. (2017) in a study using the University of Basilicata ground based Raman Lidar (BASIL) and the air-borne DIfferential Absorption Lidar (DIAL) systems, focusing on three HPEs of the autumn 2012 period. Such variations occurred in an hourly time-window when low-level wind conditions transitioned from a northerly Mistral/Tramontane flow to southerly marine. Chazette et al. (2015b) reported similar ranges for PBL moisture variation measured by the ground-based Water-vapour Raman Lidar (WALI) during the same autumn period in the Western Mediterranean (WMed).

Focusing on the moisture variations preceding heavy precipitation, past findings by Khodayar et al. (2016a) in the WMed during autumn, have demonstrated that IWV can rise up to 20*mm* before some events. In particular for an HPE of the 2012 autumn, Khodayar et al. (2018) found that the organized convective systems impacting the most the Mediterranean coasts were preceded by abrupt increases of IWV (10*mm* in a 6 h to 12 h time-window). Considering more HPEs of the 2011 and 2012 autumn seasons, Khodayar et al. (2016a), found that differences up to ca. $8mm$ exist in the diurnal cycle of IWV, over areas affected by heavy precipitation between seasonal mean values and these periods of heavy precipitation.

2.2 Atmospheric Moist Convection

Atmospheric convection is defined as the thermally directed motion of a fluid resulting from the action of gravity upon an unstable vertical distribution of mass, where "vertical" means along the gravitational vector (Emanuel, 1994). This definition understands "convective" motions as generally highly turbulent, excluding "non-convective" motions such as simple Hadley circulations or sea breezes regarded as laminar (except at the boundaries). Examples of atmospheric phenomena where convective motions take place, range from microscale turbulence to large organized precipitating systems such as squall lines or hurricanes.

In particular, precipitating convection is understood as cloud formations resulting from the vertical motions, which can show different degrees of organization, to a large extent influenced by the vertical wind shear (Markowski and Richardson, 2010). Figure 2.2 represents the spectrum of organized convective storms as a function of vertical wind shear in the 0-6 km layer. This is the dominating factor influencing the type of convection organization, although factors such as the vertical distribution of buoyancy and moisture can also be relevant

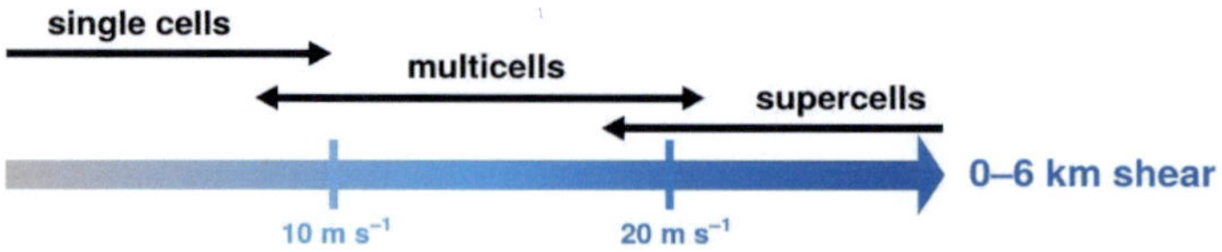

Figure 2.2: Spectrum of storm types as a function of vertical wind shear. Vertical shear exerts the greatest influence in organization type for convective events. Still other factors such as stratification of moisture or temperature also influence the organization type. Obtained from (Markowski and Richardson, 2010)

in shaping convection at the mature stage. Single cells are prone to develop in environments of weak vertical wind shear, with a lifetime of ca. $30-60\ min$. For higher 0-6km vertical wind shear (between $10\ m\ s^{-1}$ and $20\ m\ s^{-1}$) the gust front is able to initiate repeatedly new cells, this is called multicellular convection. Examples of multicellular convection are Mesoscale Convective Systems (MCSs), squall lines or Mesoscale Convective Complexes (MCC). Finally, supercellular convection takes place in large shear environments ($> 20\ m\ s^{-1}$), were the propagation is governed by vertical pressure gradients rather than gust-front lifting. In the WMed region most heavy precipitation events are brought about MCSs organizing under favourable large-scale situations which continuously trigger new cells and can last up to one day (Khodayar et al., 2015; Duffourg et al., 2016; Lee et al., 2016; Khodayar et al., 2018). Additionally, other convection types have been reported by past studies, such as, squall lines (Hally et al., 2014; Caldas-Álvarez et al., 2017) and supercells (Manzato et al., 2015; Davolio et al., 2016).
The mass and momentum conservation equations describe the movement of the fluids in the atmosphere. These movements are forced by the gravitational force ($\rho \boldsymbol{g}$), the pressure gradient force ($-\boldsymbol{\nabla} p$), the Coriolis force ($-2 \cdot \boldsymbol{\Omega} \times (\rho \boldsymbol{V})$), with $\Omega = 7.29 \cdot 10^{-5}\ rad\ s^{-1}$ and the viscosity ($\boldsymbol{F}$).

$$\rho \frac{d\boldsymbol{v}}{dt} = -2 \cdot \boldsymbol{\Omega} \times (\rho \boldsymbol{V}) - \boldsymbol{\nabla} p + \rho \boldsymbol{g} + \boldsymbol{F} \tag{2.4}$$

The horizontal and vertical directions of Eq. 2.4 can be split and the conservation of mass equation closes the relation between them. This implies that the vertical velocity at any height (z_1) can be related to the net horizontal convergence.

$$\int_0^{z_1} \left(\frac{\partial u}{\partial x} + \frac{\partial v}{\partial y} \right) dz = - \int_0^{z_1} \frac{\partial w}{\partial z} dz \tag{2.5a}$$

$$\frac{dw}{dt} = -\frac{1}{\rho} \frac{\partial p}{\partial z} - g + F_w \tag{2.5b}$$

Under the assumptions of no viscosity and a horizontally homogeneous pressure and density states, the vertical movements of mass in the atmosphere can be split into the forcing, exerted by the large-scale pressure anomalies and the buoyancy term dependent on the deviations of pressure and density from the horizontally homogeneous field. First, the Reynolds decomposition in mean and turbulent components for pressure and density is applied to Eq. 2.5b ($p = \overline{p} + p'$, $\rho = \overline{\rho} + \rho'$).

$$\rho \frac{dw}{dt} = -\frac{\partial p'}{\partial z} - \rho' g \tag{2.6a}$$

$$\frac{dw}{dt} = -\frac{1}{\rho}\frac{\partial p'}{\partial z} + B \tag{2.6b}$$

Where B is the buoyancy $B = -\rho'\rho^{-1}g$ (Markowski and Richardson, 2010). Hence, Eq. 2.6b shows that, after neglecting the viscosity term, one can obtain the equation for vertical wind speed and that it is controlled by the *non-hydrostatic pressure gradient acceleration*, arising from dynamical effects and a second term which depends on the density anomalies (Emanuel, 1994).

Moist thermodynamics

The vertical movements of the air parcels in the atmosphere depend on the density differences, the ideal gas equation relates the density of an air parcel to its temperature and pressure. Expressed as a mixture of dry air with water vapour, the ideal gas equation for moist air is.

$$p = \rho R_d T_v \approx \rho R_d \cdot T\,(1 + 0.61r) \tag{2.7}$$

Where T_v stands for the virtual temperature of the air parcel, obtained as the product of the temperature of dry air times a factor depending on the water vapour mixing ratio (r) and the ratio between the apparent molar mass of dry air and water vapour. R_d is the gas constant for dry air ($R_d = 287.04\ Jkg^{-1}K^{-1}$).

Since in the atmosphere all movements occur in a stationary pressure force field, every vertical displacement will be subject to variations of the pressure in turn affecting temperature. In order to work with a conserved quantity, the potential temperature is defined so that it is conserved under adiabatic displacements of the air parcel.

$$\theta = T\left(\frac{p_0}{p}\right)^{R_d/c_{p,d}} \tag{2.8a}$$

$$\theta_v \approx \theta\left(1 + 0.61r - r_l\right) \tag{2.8b}$$

The potential temperature θ can be defined as the temperature of an air parcel if it were compressed adiabatically to a reference pressure level (p_0), usually 1000hPa. If instead of dry air, moist air, is considered, a development analogue to that of Eq. 2.7 leads to the definition of the virtual potential temperature, θ_v (see Eq. 2.8b). Where r_l is the mixing ratio of liquid water in the air and $c_{p,d}$ is the specific heat of dry air (1005 $J\ kg^{-1}\ K^{-1}$). θ_v is conserved under adiabatic unsaturated motions of an air parcel compound of dry air and water vapour.
Further expressions for humidity variables and expressions referred to in this thesis are not included here as they are contained in most background books of atmospheric science.
In the course of a parcel ascent, phase transitions take place with the consequent latent heat exchange. The changes of phase, render θ and θ_v non-conserved quantities, and it is therefore useful defining thermodynamic coordinates conserved under phase transitions. One can define a *reversible moist adiabatic* process which is a process where the total water content (i.e. condensate and vapour) is constant (Markowski and Richardson, 2010). Many authors also use the concept of a *pseudoadiabatic* process which also allows the definition of a variable conserved under phase changes. Indeed, a *pseudoadiabatic process* is one where instead of considering the total water content constant, it is assumed to precipitate as fast as it condensates (Markowski and Richardson,

2010). For practical purposes the two approaches are used frequently without distinction.

Under the assumption of conservation of the total water content, one can define the Equivalent Potential Temperature (θ_e), which can be understood as a "condensation" temperature (Stevens, 2005). This is the temperature, in K, an air parcel would have if it were brought to the reference pressure level ($p_0 = 1000hPa$) and all water vapour had condensed.

$$\theta_e = T\left(\frac{p_0}{p_d}\right)^{R_d/(c_{p,d}+r_t c_l)} \times exp\left[\frac{l_v r_v}{c_{p,d}+r_t c_l}\right] \tag{2.9}$$

θ_e is widely used in Deep Moist Convection (DMC) studies characterizing moist reversible displacements of the air parcels after the Lifted Condensation Level (LCL) is reached and is a good proxy for buoyancy of the displacing air parcels (Honda and Kawano, 2015). l_v stands for the enthalpy of vaporization. In Eq. 2.9, p_d stands for the pressure of dry air, r_t for the hydrometeor mixing ratio (mass of the water vapour plus condensate per unit mass) and c_l for the specific heat of liquid water at constant pressure ($\approx 4200\ J\ kg^{-1}\ K^{-1}$).

Environmental lapse rate and the parcel method

Vertical anomalies of density set how buoyant a certain atmospheric layer is with respect to the layers aloft and below. These density anomalies at a certain level are a function of pressure, composition and phase changes. Given that by defining θ, θ_v and θ_e one can convert the contributions of pressure variations, moisture content and release/absorption of latent heat to temperature units, knowing their vertical distribution provides information about instability. The gradient of temperature with height is called lapse-rate and its derivation for dry air is straight forward since it stems from the hydrostatic equation and the first law of thermodynamics (Holton and Hakim, 2012).

$$\Gamma_d = -\frac{dT}{dz} = \frac{g}{c_{p,d}} \approx 9.8\ K\ km^{-1} \tag{2.10}$$

This implies that dry air rising cools ideally at a rate of $9.8K$ per km of ascent. When the water vapour condensates, the release of latent heat is to be taken into account. The air parcel does not follow the dry-adiabat anymore but instead a reversible moist adiabat of lapse rate.

$$\Gamma_m = -\frac{dT}{dz} = \frac{g}{c_{p,d} + r_t c_l} + \frac{1}{c_{p,d} + r_t c_l}\frac{dl_v r}{dz} \tag{2.11}$$

Where $c_{p,d}$ and c_l are the specific heat capacity of dry air and liquid water, at constant pressure ($c_l \approx 4200\ Jkg^{-1}K^{-1}$), respectively and r_t and r are the mixing ratio of total water (water vapour and condensate per unit mass, $r_t = r + r_l$) and water vapour, respectively. The process of condensation imposes $d\left(l_v r\right)/dz < 0$, hence is $\Gamma_m < \Gamma_d$, indicating that moist air will cool down at a lower rate than dry air.

Especially useful in the study of buoyancy, lapse-rate and convective ascents, is the use of thermodynamic diagrams. In this thesis one of them is frequently used, the Skew-T vs. log-P diagram, specially suited for the study of DMC as several parameters (significant levels, Convective Available Potential Energy (CAPE) and Convective Inhibition (CIN), lapse-rates) are simple to represent for each situation. Figure 2.3 shows a schematic of this diagram with an example profile denoted by the solid black line for Temperature (T) and by the dashed line for Dew Point Temperature (T_d). The skewed vertical isolines stand for isotherms whereas the horizontal parallel isolines denote isobars. The straight dashed thin black line stands for the value of the saturation mixing ratio of the parcels close to the surface. Additionally, the dry adiabatic lapse rate (solid thin black line) and the reversible moist adiabat are shown (curved dashed black lines).

The Skew-T log-P diagrams are useful to quantify the CAPE and CIN (Moncrieff and Miller, 1976) of a certain vertical atmospheric distribution using the *lifted parcel method* (Leuenberger et al., 2010). In the parcel method, an atmospheric parcel close to the surface is theoretically lifted, first dry-adiabatically (before condensation) and then in a moist reversible process (after condensation). This method assumes that convective motions of a parcel rising from the surface happen fast enough to consider them reversible and therefore no heat exchange occurs with the environment (i.e. the process is adiabatic). When this parcel is lifted three are the levels relevant for its evolution, the LCL, the Level of Free Convection (LFC) and the Equilibrium Level (EL), see Fig. 2.3. These are respectively, the height at which the lifted air parcel endures condensation, second, the level at which it becomes buoyant (with respect to the surrounding

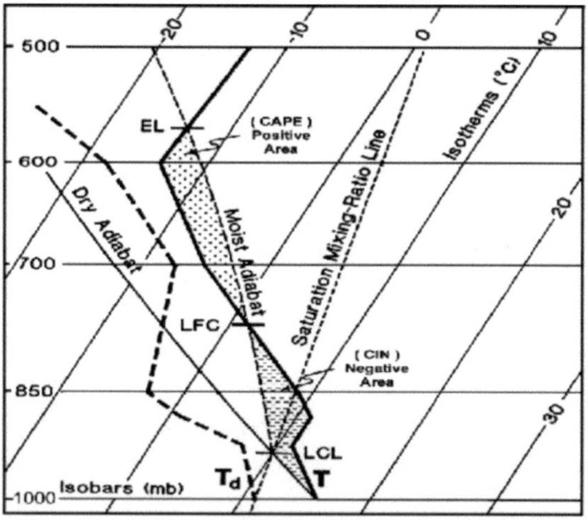

Figure 2.3: Schematic of a Skew-T vs. Log-P diagram. The T_d and T of an example profile are shown as thick black lines, dashed and straight, respectively. The dry and moist adiabats, are represented by a leaning thin black line and an arched thin dashed black line, respectively. The three convection levels are highlighted and the areas between the adiabats and the example environmental profile (jagged thick black line) denote CAPE and CIN. The skewed thin black lines represent the isotherms and the parallel horizontal lines the isobars. Skew-T vs. log-P, obtained from *www.stuffintheair.com*, last accessed on 10-Nov-2018.

environment) and finally, the level where buoyancy is inverted and the lifted air parcel cannot ascend any further. CAPE and CIN are calculated considering the buoyancy of the lifted parcel integrated in the vertical direction. In this thesis, the parcel selected for lifting is a mean of the lowest 50 hPa. CAPE and CIN are obtained as follows.

$$CAPE \approx g \int_{LFC}^{EL} \frac{T_v^p - T_v^e}{T_v^e} dz \qquad (2.12a)$$

$$CIN \approx -g \int_{0}^{LFC} \frac{T_v^p - T_v^e}{T_v^e} dz \qquad (2.12b)$$

Where T_v^p and T_v^e are the virtual temperature of the lifted parcel, and of the measured environmental profile, respectively. The limits of the integrals are the three convective levels, where CIN is integrated between the ground and the first point of buoyancy reversal (LFC), and CAPE is integrated between that level and the EL. In order to build up the profile of the theoretically lifted parcel, one needs to follow the dry-adiabat starting close to the surface (in Fig. 2.3 is at 1000 hPa). After reaching the LCL, denoted in the diagram as the intersection between the saturation mixing ratio and the dry adiabat, the air parcel continues its ascent following the moist adiabat. At the point where the virtual temperature of the air parcel is larger than that of the measured environmental profile, it becomes buoyant, this is the LFC. Then the rise continues until buoyancy is again inverted at the EL.

Instability

The parcel method has been introduced, which quantifies the instability of an air parcel to finite amplitude displacements in the vertical direction. This conceptual framework for instability is complementary to that of instability to infinitesimal displacements of the air masses at a specific atmospheric height. Both conceptions are commonly used in literature.

Instability under finite amplitude displacements is understood as the difference between an observed lapse-rate (Γ_e) and that of a theoretically lifted air parcel following first a dry adiabatic (Γ_d) and then a reversible moist adiabatic (Γ_m) process. In this regard, three situations are possible

- $\Gamma_e < \Gamma_m < \Gamma_d$ - Absolutely stable
  The buoyancy of the lifted air parcel in a finite displacement is negative with respect to its environment, regardless of whether the lifted parcel is rising dry or moist adiabatically. Hence, further rising is not promoted by the buoyancy forces

- $\Gamma_e > \Gamma_d > \Gamma_m$ - Absolutely unstable
  The lifted air parcel is positively buoyant in any case and the lifted air parcel is unstable in dry and moist adiabatic states.

- $\Gamma_m < \Gamma_e < \Gamma_d$ - Conditionally unstable
  At this layer of the atmosphere, the lifted parcel is stable for unsaturated conditions but unstable if condensation takes place.

The parcel method uses the concept of *conditional instability* to describe qualitatively unstable situations, where CAPE is used for quantification.
On the other hand ***Instability under infinitesimal displacements*** uses the concept of *potential instability* to describe qualitatively unstable situations and the KO-index (see further in this section) for quantification. Potential instability is introduced by Emanuel (1994) "Were the entire air mass lifted bodily until it becomes saturated, then the upward decrease of θ_e does imply instability". Hence for potential instability to be present, saturation must occur and $d\theta_e/dz < 0$. If a certain air parcel is displaced infinitesimally to a state $\theta_e^{p'}$, the relative comparison against the equivalent potential temperature of the environment (θ_e^e) will define the instability conditions.

- $\theta_e^{p'} < \theta_e^e$ - Stable parcel
  The air parcel infinitesimally displaced, will go back to its original position as the buoyancy force imposes equilibrium

- $\theta_e^{p'} \approx \theta_e^e$ - Neutral parcel
  After the infinitesimal displacement, the air parcel will stay in that position since no buoyancy force is available for restoring the former position.

- $\theta_e^{p'} > \theta_e^e$ - Unstable parcel
  The air parcel is not in equilibrium with its environment and after imposing an infinitesimal vertical motion, the air parcel will continue to rise.

Just as CAPE gives a quantification of conditional instability, the degree of potential instability can be quantified by the use of another convective index, the KO-index (Andersson et al., 1989).

$$KO = \frac{1}{2}\left(\theta_e^{700} + \theta_e^{500} - \theta_e^{1000} - \theta_e^{850}\right) \quad (2.13a) \qquad KO = \frac{1}{2}\left(\theta_e^{700} + \theta_e^{500} - 2\theta_e^{850}\right) \quad (2.13b)$$

KO has units of $[K]$ and the definition of Eq. 2.13bb is used for situations of low surface pressure where no values of θ_e are available at 1000hPa. The superscripts of θ_e indicate the pressure levels at which these are to be evaluated. KO index is especially suited for evaluating instability in situations of strong large-scale forcing. This is the main advantage of KO-index over CAPE since the latter has a strong dependence on the state of the parcel close to surface selected for lifting. On the other hand, KO considers the stratification of θ_e up to the 500 hPa level, and hence it contains as well information on how favourable the upper levels are for lifting. KO index is used to determine situations of unstable stratification associated with potential for storm development. The classification is as follows.

$KO > 6$	**Potentially stable** No potential for storm development
$6 \geq KO > 2$	**Potentially indifferent** Potential for isolated storms
$KO \leq 2$	**Potentially unstable** Numerous convective storms can be expected

Entrainment/detrainment processes

So far, all the introduced relations assumed adiabatic processes while the air parcels rise. While useful for understanding the main processes of rising convective plumes, these motions cannot be considered as closed systems in reality. Mixing through in the cloud occurs in entrainment/detrainment processes.

$$E_u = E_u^T + E_u^D \tag{2.14a}$$

$$D_u = D_u^T + D_u^D \tag{2.14b}$$

Entrainment (E) in a convective plume is defined as the mixing of environmental air into the organized flow. Detrainment (D) is the flow from the organized plume to the environment. Mixing through the cloud walls, occurs in uprising and subsiding air masses. The upward directed air is affected by entrainment and detrainment from a turbulent and a dynamic contribution (see Eqs. 2.14a and 2.14b). Where the upper scripts T and D stand for Turbulent and Dynamic and the subscripts for upward (u) motions. The turbulent entrainment/detrainment coefficients are modelled as proportional to the total mass being lifted or subsiding, multiplied by a factor in the order of $10^{-4}m^{-1}$. Regarding the dynamic terms for E and D, the dynamic entrainment term is a function of the grid-scale local moisture convergence (Doms et al., 2011), in which convergent dry air will be detrimental to the buoyancy of the plume (see

Eq. 2.15a). In the model used in this thesis, the Consortium for Small-scale Modelling (COSMO) model, the dynamic detrainment is simulated to occur only in the upper regions of cumulus clouds, where the buoyancy force inverts its sign and does not promote further the rising of the air masses. These are the levels where convective motions start to decelerate, i.e. passed the EL.

$$E_u^D = -\frac{\rho}{q}\boldsymbol{V}\cdot\boldsymbol{\nabla} q \tag{2.15a}$$

$$D_u^D = \begin{cases} (1-b_u)\,(M_u)_{k+1/2}\,/\Delta z_k & if\ f = k_T \\ b_u\,(M_u)_{k+1/2} & if\ f = k_T - 1 \\ 0 & else \end{cases} \tag{2.15b}$$

Where ρ, q and $\boldsymbol{V}$ stand for the density of the air, the specific humidity and the wind vector, respectively. b_u is a parameter set artificially, to ensure that only a fraction of the updraft mass flux is available for lateral outflow at the k_T and $k_T - 1$ levels. M_u is the mass updraft and Δz_k is the distance between atmospheric levels.

Adiabatic and evaporative cooling within the updrafts

The condensation/evaporation processes endured by the rising parcels render the behaviour of the fluid non-linear (Stevens, 2005). One direct consequence of attaining saturation by the rising parcels is the release of latent heat, which, as demonstrated by the changes in the lapse rate, heats up the mixture of air. This effect promotes uprising, by increasing the buoyancy of the air parcels after the LCL is reached. This process is called *saturation by adiabatic expansion* and is by far the most important contributor to the total saturation in the atmosphere (Emanuel, 1994). This mechanism takes place, for instance, under free convective ascents, large-scale destabilization or forced ascents over topography. As the air rises, it cools and the saturation vapour pressure falls setting a critical state (T^*, p^*, r^*) beyond which condensation starts. From the definition of relative humidity, the Clausius-Clapeyron equation and the first law of thermodynamics the critical temperature at which saturation will take place, is defined (T^*). The heat released to the mixture depends on the latent heat of condensation/vaporization ($l_v r$). Recalling e is the vapour pressure

$$T^* = \frac{2840}{3.5\ ln\ T - ln\ e - 4.805} + 55 \tag{2.16}$$

The opposite mechanism may also take place, once precipitation is already present, i.e. evaporation of precipitating rain (or ice). This process is known as *evaporative cooling* (Emanuel, 1994; Doswell et al., 1996). This has a direct effect in reducing buoyancy within the updraft, given the absorption of latent heat to evaporate the rain or ice. This effect adds to the dynamical effect of buoyancy reduction due to the downwards drag of condensed water. *Evaporative cooling* is realized in an isobaric process in which the saturation vapour decreases and the heat of vaporization is supplied by the environment. If this process continues long enough it ideally approaches saturation. The temperature attained by the air mass is called the wet bulb temperature (T_w), defined from the enthalpy equation for an isobaric process.

$$\left(c_{p,d} + r^*\left(T_w\right) c_l\right)\left(T_w - T\right) + l_v\left(T_w\right) r^*\left(T_w\right) - l_v\left(T\right) r = 0 \tag{2.17}$$

After the wet bulb temperature is reached, the air parcels continue their descent along a pseudoadiabat. It must be noted that this is an ideal explanation of the process and that irreversible and non-linear processes are involved.

Soil moisture and surface exchanges

Given the strong dependence of convection on thermodynamic effects, the role of the soil (and vegetation) in inputting and absorbing energy and moisture to the atmospheric system needs to be addressed. The link between soil effects and precipitation has been the object of study of numerous research studies

(Seneviratne et al., 2010). Beyond the intuitive understanding of a moister soil inducing larger precipitation, indicating a positive feedback, these studies have demonstrated that the indirect interactions are prominent actors in convection, that can at times lead to a precipitation reduction. For example Hohenegger et al. (2009), reported a dry soil advantage for precipitation in Convection-Permitting Model (CPM) simulations over the Alps, given drier initial soil moisture yielded more vigorous thermals which could more easily break the stable air inversions.
At the soil surface, the water and heat balances can be expressed as follows.

$$\frac{dS}{dt} = P - E - R_s - R_g \tag{2.18a}$$

$$\frac{dH}{dt} = R_n - l_v E - SH - G \tag{2.18b}$$

Where dS/dt is the change of water content within the given layer, P is the precipitation, E is the evapotranspiration, R_s is the surface runoff, and R_g is the drainage (Seneviratne et al., 2010). Regarding the change of energy (dH/dt), R_n is the net radiation, λE is the latent heat flux, SH is the Sensible Heat flux and G is the ground heat flux to deeper layers.
The soil moisture will provide water vapour and energy (as latent heat flux, through the evaporation term (E) which is to be added to the Sensible Heat (SH). The soil moisture controls the partitioning of sensible and latent heat fluxes. Recent modelling studies with COSMO on the sensitivity to soil moisture changes (Barthlott and Kalthoff, 2011; Gantner et al., 2017) have reported a compensating effect in the outgoing latent and sensible heat fluxes, so that in the case of the former showing an increase, the latter shows a decrease.

Orographic effects on precipitating convection

Orographic elevations are relevant for atmospheric convection since the dynamic forcing of the orography may eventually give aid in overcoming any possible stable layers. Three factors explain the spatial distribution of precipitation near or over an orographic feature (Houze, 2012), the microphysics of the particle growth, the dynamical behaviour of the fluid flow and the thermodynamics of moist air. The first one is crucial to determine whether the hydrometeors will grow sufficiently to precipitate over the windward side of the mountain, at the lee side or none of them.

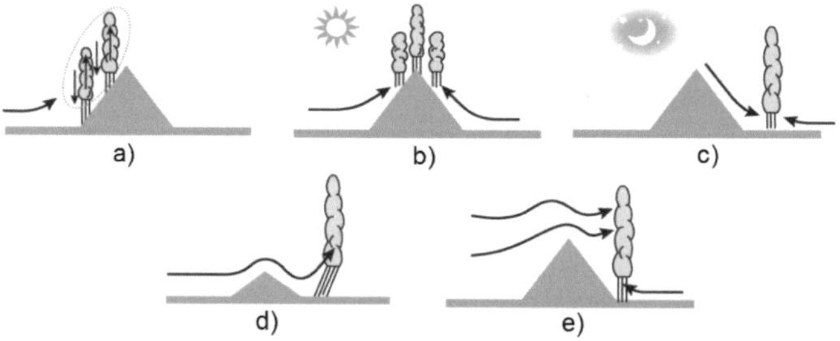

Figure 2.4: Mechanisms by which mountain and hills affect convective precipitation. a) Example of convective cells triggering due to dynamic lifting at the mountain foothills. b) Daytime convergence at the mountaintop due to solar radiation effects. c) Night-time convection at the foothills arising from convergence due to downslope winds. d) Example of convection generation by lee-side waves. e) Convection induced by wave propagation.

This depends on the rimming rate of the hydrometeors, the chemical composition of the aerosol and the height and shape of the orographic feature. Regarding the dynamics of orographic flow, convection will trigger if the atmospheric stratification becomes buoyantly unstable after lifting. Finally, the thermodynamics of moist air set the height at which saturation of water vapour is reached (LCL), the height of the LFC level and the height of any possible inversions. Figure 2.4 shows five mechanisms of the orography affecting or bringing DMC. The first mechanism, Fig. 2.4.a, represents the situation of orographic lifting aiding the low-level air masses to reach the LFC. This situation may occur under four scenarios. First, if a deep layer of very unstable air is lifted beyond the LFC; second, if buoyant instability is contained in a shallow layer at low levels, whence convection is limited to the foothills; third, convection triggers embedded in a pre-existing cloud system and fourth, in the presence of strong shear in the air flowing over the elevated terrain. Figures 2.4.b) and 2.4.c) represent the convection related to the diurnal heating cycle, where at day-time, convection forms at the mountaintops, aided by upslope wind flows and intense diurnal heating at the top, where upward winds converge. During nighttime the flow is opposite and blows downslope where, in the case of converging with low-level moist air masses, convection is also likely to develop. Figure 2.4.d) indicates how lee-side waves generate convection and Fig. 2.4.e, is an example of convection caused by a vertically propagating wave motion (Houze, 2012).

Degree of synoptic forcing

The analysis of the characteristic time scales of convection leads to defining two different regimes of convective phenomena.

Type 1 - - **Weak synoptic forcing** - The time scale of CAPE creation by the large-scale processes is long compared to that needed by convection to stabilize the atmosphere.

Type 2 - - **Strong synoptic forcing** - CAPE consumption by the convection balances CAPE creation by the large-scale processes.

Type 1 convection is linked to how, where and when a lifting mechanism helps to overcome any existing CIN and depends to a large extent on the mesoscale factors to trigger and be maintained. Convection Type 2, is the form of the global convection as a response of the fluid to destabilization imposed by large-scale ascent. In these situations, the duration of convection is typically shorter than large-scale synoptic time-scales and convection is said to be in a state of statistical equilibrium with its forcing (*statistical equilibrium convection*).
Provided, this classification of degrees of synoptic forcing in terms of the time-scale of removal of instability by convection, quantifying it through a parameter is useful to differentiate between regimes. Following Keil et al. (2013) and Molini et al. (2011), the Convective Adjustment time-scale (τ) is defined as

$$\tau = \frac{CAPE}{dCAPE/dt} = 0.5 \cdot \left(\frac{\rho_0 c_p T_0}{l_v g}\right) \frac{CAPE}{P} \tag{2.19}$$

Where τ is the convection adjustment time scale and the rate of change of CAPE ($dCAPE/dt$) can be can be expressed in terms of the vertically integrated latent heat release, that, in turn, can directly be determined from the precipitation rate (P in $[kg\ s^{-1}\ m^{-2}]$), the latent heat of vaporization (l_v), the reference values for density ($\rho_0 = 1.2\ kg\ m^{-3}$) and temperature, temperature ($T_0 = 300\ K$), the gravity (g), and the specific heat at constant pressure ($c_p = 1005\ J\ kg^{-1}\ K^{-1}$) as in Keil et al. (2013).
The selection of the thresholds for τ stems from the idea that convection forced by large-scale processes removes CAPE fast, compared to their characteristic time-scale (of ca. 12h). On the other case type, if convection does not remove CAPE fast enough, τ will be larger than the large-scale systems characteristic time. Molini et al. (2011) set this threshold to be of $< 6\ h$ for cases of strong synoptic forcing and $> 6\ h$ for weak synoptic forcing whereas Keil et al. (2013) chose $\tau < 3\ h$ for convection of type 2 (strongly forced) and $\tau > 12\ h$ for convection of type 1 (weakly forced). The somewhat arbitrary selection of the τ thresholds is due to the fact that these two regimes are at the tails of a continuous spectrum. In other words that DMC may as well be arising from a combination of favourable mesoscale factors and supporting large-scale dynamics.

2.3 Atmospheric Moisture Impact on Deep Moist Convection

Moisture is a crucial element of convection initiation, maintenance and termination. Doswell et al. (1996) enumerated three ingredients needed for DMC development. A conditionally unstable environmental lapse-rate, sufficient moisture and some mesoscale mechanism by which the air parcels are lifted to the LFC. Examples of dynamic forcing for convection initiation are upper-level lifting, low-level convergence, or the build-up of cold pools initiating new convective cells.

High IWV areas have been located upstream of regions affected by heavy precipitation (Khodayar et al., 2016a; Lee et al., 2018), demonstrating the link between these two quantities. For instance, Khodayar et al. (2018) found for a case study of heavy precipitation regions of 30 mm to 45 mm in IWV upstream of the Gulf of Liguria and Sicily, locations affected by heavy precipitating MCSs. High atmospheric moisture amount is related to convection initiation, as introduced by Sherwood et al. (2010). The storms initiate along convergence lines showing large boundary layer moisture and surface layer moisture variations, as well as vertical moisture variations and of the depth of the moisture distribution.

A very moist low-level layer increases the probability of convection triggering (Crook, 1996; Li and Shen, 2010; Chazette et al., 2015b) and its intensity given it lowers the LFC (aiding CIN overcoming) and enhances CAPE. With higher moisture at the lower troposphere the buoyancy of the lifted air parcel tends to be larger than that of the environment and as in a *locomotive* effect, the larger number of water vapour droplets releases more latent heat to promote the uprising. Seasonal, as well as event-scale studies, have shown the link between high moisture content, high CAPE and convection. Khodayar et al. (2016a) assessed the excess in CAPE in the diurnal cycle to be 1000 $J\ kg^{-1}$ larger during heavy precipitation during the autumn period. In the analysis of two HPEs of the autumn 2012 period in the WMed, Khodayar et al. (2018) and Khodayar et al. (2015), assessed maximum values of CAPE up to 2000 $g\ kg^{-1}$ during the befalling of heavy precipitation.

An identified mechanism for PBL moistening are the convective rolls (counter-rotating horizontal vortices over uniform terrain under typically moderate surface heat fluxes and low wind shear conditions). A series of studies (Weckwerth et al., 1996; Weckwerth, 2000; Huang et al., 2009) have addressed the role of convective rolls in moistening the PBL and favouring convection. One the main findings of these studies is that the updraft regions of the rolls promote larger moisture (1.5 $g\ kg^{-1}$ larger) throughout the whole PBL at the locations of the observed cloud bases. Hence are the updraft regions of the convective roles related to the location of convective development. Given the wavelength of the convective roles is between 3 *km* to 5 *km*, rarely does a radiosonde network have a sufficient spatial coverage to sample the updraft regions.

Besides moistening at the lowest levels, moistening at the lower free troposphere is relevant for the depth, duration and intensity of convection. Moistening above the PBL has been pointed out as a mechanism supporting the development of organized convection (Hohenegger and Stevens, 2013; Chazette et al., 2015a). Given, mixing exists through the walls of the convective plume, i.e. entrainment and detrainment processes (see Section 2.2), a too dry lower free troposphere (850 *hPa* to 700 *hPa*) will reduce the buoyancy of the rising plume. This modifies the depth of convection and can even terminate it if for example cold dry air is intruded at those levels (Honda and Kawano, 2015; Khodayar et al., 2015). Besides, the mixing through the wall clouds. The free-atmospheric moisture will influence the condensation evaporation mechanisms by changes of the water vapour concentration. However more advancement in the knowledge of the role of moisture above the boundary layer is needed since high humidity at those levels is not always beneficial for convection growth (Sherwood et al., 2010). Hohenegger and Stevens (2013), addressed this issue by studying whether congestus clouds moistened the lower free troposphere enhancing convection. They concluded that the time-scale of the transition from congestus to deep convection is too fast (2 *h* over land 4 *h* over the ocean) in comparison to the time needed by congestus cloud to sufficiently moisten the atmosphere (10 *h*).

These impacts on instability and saturation of the convective cloud demonstrate the relevance of the thermodynamic stratification of the atmosphere, especially at the PBL and at the lower limit of the free troposphere. The sensitivity of convection to the vertical distribution of moisture in the atmosphere has been investigated by past studies utilizing numerical weather and climate models. Sensitivity experiments in realistic set-ups, have revealed very large effects on precipitation amount. Crook (1996) demonstrated that differences of 1 $g\ kg^{-1}$ in the humidity gradient between the surface and the well-mixed boundary layer, make the difference between representing no convection whatsoever and intense convection. Keil et al. (2008) showed that if increased by 30% at the PBL, humidity brings an intensification of precipitation with totals being five times larger. Honda and Kawano (2015) reduced relative humidity by 10% at the lower limit of the free troposphere in the environment of a developing supercell storm using the Advanced Regional Prediction System (ARPS) model. They reported an appearing dry-air penetration near the surface, resulting in a reduction of the parcel buoyancy and the suppression of deep convection. In another modelling study, simulating the back-building Mesoscale Convective Systems (MCSs) of an HPE in the autumn 2012 period, Lee et al. (2018), quantified the precipitation reduction to be of 10% under drying of the atmosphere at the surface to 1km deep layer.

3 Heavy Precipitation in the Mediterranean

Heavy precipitation causes large monetary losses and casualties due to their related hazards in the Western Mediterranean (WMed). Previous studies (Sala et al., 2000; Drobinski et al., 2016; Tramblay and Somot, 2018; Alfieri et al., 2018) have warned about the effect of Climate Change in the Mediterranean basin regarding the intensification of heavy precipitation events and their frequency. The Mediterranean region is thus considered a hot spot for heavy precipitation given the combination of several characteristic factors. These factors are, it is in the transition zone from sub-tropical to temperate climates, it is an almost enclosed basin, it is surrounded by steep orographic areas and that by autumn shows a especially high Sea Surface Temperature (SST). Given the relatively high SSTs by autumn, the Mediterranean Sea is a heat and moisture reservoir for heavy precipitation and the whole basin is specially propitious to atmospheric situations where very moist and warm low level flows coincide northern cold air deepening down to mid-latitudes.

This chapter summarizes the main features of the Mediterranean climate (regional climate types, orography, sea conditions and dominant wind circulations) and the occurring synoptic and mesoscale situations in autumn leading to heavy precipitation. Additionally, given the many resources used from its scientific network, the HYdrological cycle in the Mediterranean eXperiment (HyMeX) is presented. Moreover, the unique field campaign of the HyMeX initiative, the Special Observation Period 1 (SOP1), provides the radiosondes and rain gauges observational data used in this thesis. The HyMeX network also provides the GPS-derived column atmospheric moisture data set used in the moisture sensitivity experiments.

3.1 Characteristics of the Mediterranean Basin Conducive to Heavy Precipitation

The western Mediterranean region has a broad range of climate types given it is located between the subtropical and the temperate zones, it has mountain regions (snow climates) and even has areas of great aridity (Lionello, 2012). Hence the complexity of assessing its overall climate system. A general definition of Mediterranean climate is given by Köppen (1900) "mid-latitude temperate climate with a dry summer season, which can be either warm or hot". An extended definition of the Mediterranean climate is given by Bolle (2003), a "temperate rainy, humid meso-thermal climate with dry subtropical warm to hot summers". The Figure 3.1, shows the classic classification of the climate systems by Köppen (1900) for the Mediterranean basin. South of the basin, over the Maghreb area and the Arabian Peninsula, the predominant climate type is desert, with the only exceptions of northern Morocco at the strait of Gibraltar and a narrow band of elevated terrain across northern Algeria and Tunisia with a Mediterranean climate with hot summer. The southern peninsulas of Iberia, Italy, the Balkans and Anatolia show the largest variety of climate types. Over Iberia, the climate ranges from a Mediterranean climate with hot or warm summer (northern Portugal and northwestern Spain) to maritime temperate at the north. This is the predominant climate type over northern Europe. This is also the case for France, only differing along the southern coast over Corsica and over the Alps which show a humid subtropical climate and tundra, respectively. The Italian coast shows mostly a humid subtropical climate, with the exception of Calabria, Sicily, Puglia (southern Italy) and Sardinia which show a Mediterranean with hot summer climate. A large contrast exists, with respect to the north due to the Alps with a snow climate. The coastal and southern parts of the Balkan peninsula have parts of both Mediterranean hot

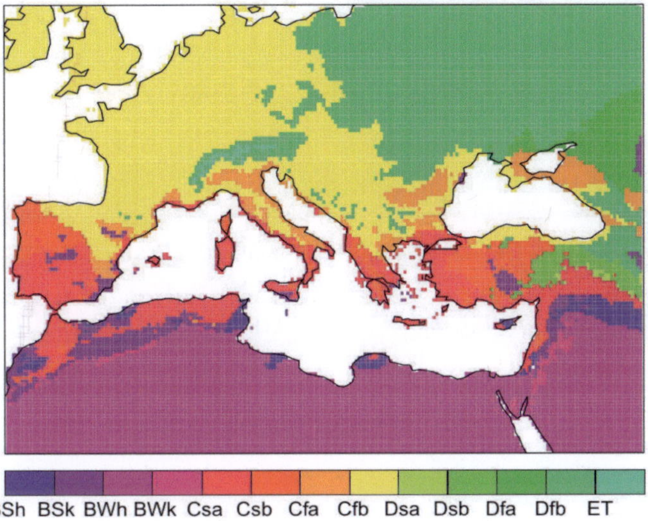

Figure 3.1: Köppen climate systems in the Mediterranean region: subtropical steppe (BSh), mid-latitude steppe (BSk), subtropical desert (BWh), mid-latitude desert (BWk), Mediterranean climate with hot/warm summer (Csa/b), humid subtropical with no dry season (Cfa), maritime temperate (Cfb), humid continental with hot/warm summer (Dfa/b), continental with dry hot/warm summer (Dsa/b), and tundra (ET). Obtained from (Lionello, 2012).

summer and humid subtropical climates, whereas the north of the peninsula shows a European maritime temperate climate.

There is, besides, a polarization of the time for the rainy seasons between the west and the east, with heavy precipitation, occurring during autumn over the northwestern part of the basin and during late winter for the south-eastern parts (Raveh-Rubin and Wernli, 2015). This can be explained by the strong link of the northern part of the basin to the North Atlantic Oscillation (NAO) and the Scandinavian Pattern (SCAND) teleconnection patterns, and by the fact that the south-eastern is connected to the descending branch of the Hadley cell (Lionello, 2012).The Mediterranean region is also influenced by tropical and subtropical systems such as tropical cyclones, events of Saharan dust and the El

Niño Southern Oscillation (ENSO) with variable impacts in heavy precipitation (Drobinski et al., 2007).

The annual cycle of precipitation in the Mediterranean is characterized by dry summers frequently associated with very long drought periods, followed by autumn and winter precipitation that are very intense (Drobinski et al., 2007). The climatology of heavy precipitation in the WMed region, shows that events with daily precipitation rates larger than 200mm are not uncommon during autumn (Drobinski et al., 2007). Regions prone to suffering heavy precipitation during that season are the Gulf of Lions, the eastern Spanish coast, northern Italy, the western coast of the Balkan Peninsula and western Turkey. This suggests a relevant role of the relatively high SST, which allows a larger water vapour loading and evaporation occurring at key spots of intense sea-air exchanges, associated with very strong winds, such as Mistral and Bora. Mistral is a northwesterly wind blowing into the Gulf of Lions, descending from the Rhône valley and Bora is a north-easterly wind, originating in the Balkans blowing into the Adriatic Sea.

Regarding future projections of heavy precipitation over the Mediterranean basin, several studies have stated the tendency for more frequent events in the warming Mediterranean climate (Sala et al., 2000; Drobinski et al., 2016; Vautard et al., 2014; IPCC, 2018). For example Tramblay and Somot (2018), analysed the relative changes in the 20-year return period (i.e., extreme precipitation that is likely to occur on average once every 20 years) between 2100 and the year of the beginning of significant heavy precipitation trend. They found a strong convergence from all models and all Climate Change scenarios showing an increase in heavy precipitation in the northern basin. Figure 3.2 shows the case for an RCP4.5 scenario. This is the scenario assuming gas emissions peaking by 2040 and then declining. This scaling of heavy precipitation with warming temperatures can be explained through the Clausius-Clapeyron equation, and the increase of water vapour content for warmer air masses.

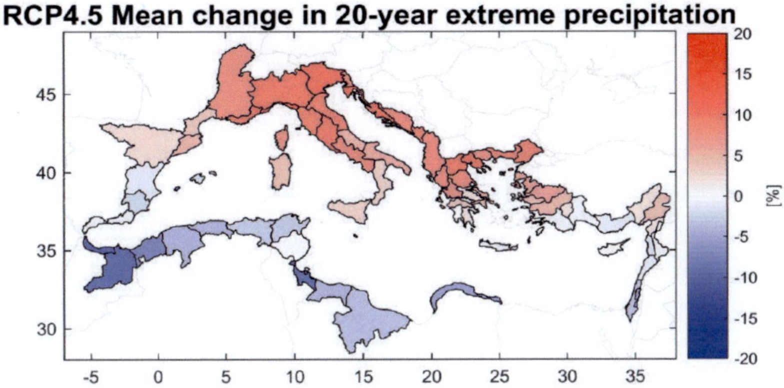

Figure 3.2: Mean relative changes between the beginning of the trend for climate change and the year 2100 in the 20-year return period of extreme precipitation under the RCP4.5 scenario. The 20-year return period is the extreme precipitation that is likely to occur on average once every 20 years. The trend for climate change is identified statistically from an ensemble of 11 members of Euro-Coordinated Regional Climate Downscaling Experiment (CORDEX) climate simulations. Obtained from Tramblay and Somot (2018).

$$\frac{de_s}{dT} = \frac{l_v e_s}{R_v T^2} \tag{3.1}$$

Where e_s is the saturation vapour pressure, l_v is the vaporization heat, R_v is the gas constant of water vapour and T is temperature. The saturation vapour pressure is larger for warmer air masses, hence these contain more water vapour which brings more precipitation events. However, Drobinski et al. (2016) demonstrated that there is a temperature change break, beyond which extreme precipitation does not scale any longer with temperature. This occurs once the moisture sources in the Mediterranean are not able to provide sufficient moisture to the precipitation systems in a too dry climate. This behaviour at high temperatures demonstrates that the mechanisms of heavy precipitation are complex and more research is needed to understand the possible impacts in a warming climate.

Synoptic and mesoscale factors inducing heavy precipitation in the WMed

Autumn synoptic situations conducive to heavy precipitation in the WMed have been extensively investigated (Khodayar et al., 2016a). Heavy precipitation occurs frequently in situations of low-pressure systems such as upper-level troughs or cut-off lows providing upper-level disturbance favouring lifting, together with a strong low-level jet, feeding moisture and locally forcing dynamic lifting (Drobinski et al., 2007; Lionello, 2012; Ricard et al., 2012; Khodayar et al., 2016a). Toreti et al. (2010) analysed the climatology of the surface pressure and geopotential height at 500hPa (Z500) during heavy precipitation in the WMed. The authors found three significant large-scale configurations anomalous to fair weather conditions, all of which, have an anomalous mid-tropospheric southwesterly flow in common. The first configuration (Fig.

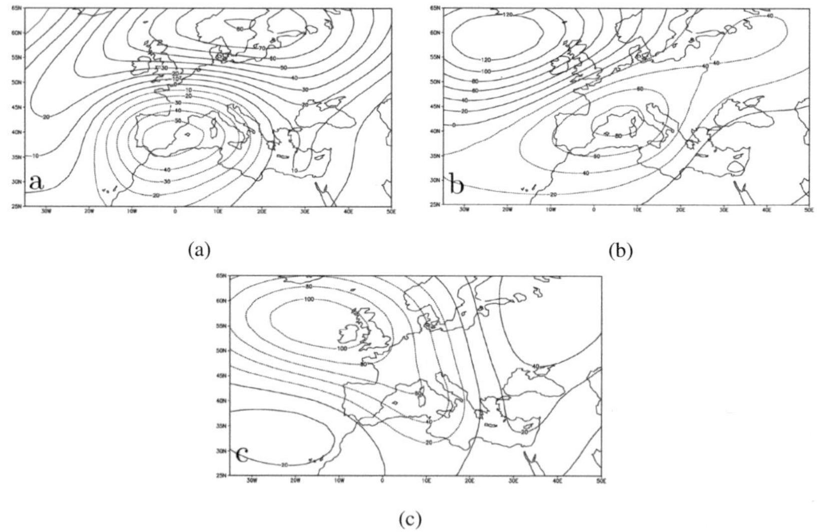

Figure 3.3: Anomalies of geopotential height at 500hPa associated with extreme precipitation in the WMed as compared to fair-weather conditions. Obtained from Toreti et al. (2010).

3.3a) was a dipole structure of Z500 with the positive anomaly placed over the Scandinavian peninsula, and the negative over the Iberian peninsula. The second also exhibited a dipole, where the positive anomaly is placed northwest to the British Isles and the negative centred at the Gulf of Lions (Fig. 3.3b). Finally, the last configuration shows (Fig. 3.3c) a triple-centre structure with a mid-tropospheric trough extending from the British Isles towards northern Africa and two anomalous positive centres in the subtropical North Atlantic and across north-eastern Europe (Toreti et al., 2010). These pressure spatial distributions stem often from descending polar air, at the change of season, from summer to winter, and from the descent in the jet stream affecting the storm track which frequently advects former tropical cyclones to the Mediterranean and north Atlantic Europe.

Although the synoptic situations associated with heavy precipitation show typically a good predictability by the models, accurately forecasting where and when precisely convection will occur is far from accomplished. This is due to the role of the mesoscale factors, which introduce uncertainty in the numerical predictions. The location and timing of moisture and instability, are linked to where and when convection will trigger. Ricard et al. (2012) pointed out, in a climatology study of the mesoscale factors related to heavy precipitation, that differences of only 200km in the location of the precipitation maximum arise from different low-level moisture flux paths. The same study stressed out that HPEs affecting southern France are commonly preceded by an intense moisture flux band extending from the eastern Spanish Mediterranean to the Avignon valley, and/or by large moisture presence between Tunisia and north-eastern Italy. The moisture flux peaks to 250 $kg\ m^{-1}\ s^{-1}$ at the mature stage of convection and is supported by Convective Available Potential Energy (CAPE) values of ca. 2000 $J\ kg^{-1}$ upstream, over the Mediterranean. An analogue result was found by Khodayar et al. (2016a) who quantified Integrated Water vapour (IWV) and CAPE mean values during the mature stage of convection during the autumn seasons of 2011 and 2012 to be close to $25-30$ mm and 1000 $J\ kg^{-1}$. A remarkable finding in Khodayar et al. (2016a), was that the

IWV and CAPE build up happens with a different timing. IWV was observed to increase in a time window of 6 h to 24 h before the onset of precipitation convection, whereas the CAPE increase is observed mainly during the hours of the convective outbreak.

In order to learn more about the moisture transported to the areas affected by heavy precipitation, Duffourg and Ducrocq (2011) analysed ten HPEs impacting the French Mediterranean during autumn seasons. They found out that the moist layer feeding the systems was confined in a thin 1000m deep layer above the sea. The origin of this moisture was assessed to be frequently at the Mediterranean Sea and, at times, over north Africa and the tropical Atlantic Ocean. The same moisture sources for HPE in the Mediterranean basin were highlighted by Winschall et al. (2014) in a statistical study using the 50 heaviest precipitation event of each season in the last decades. During summer, evapotranspiration over European land was characterized as an important moisture source whereas the North Atlantic plays the major role during autumn and Winter events with moisture uptake taking place about two days prior to precipitation onset. Additionally, tropical moisture transports were behind some HPEs in the Mediterranean, in which case those transports lasted for more than a week before reaching the areas of heavy precipitation. Once more, Krichak et al. (2014), highlights the North Atlantic Ocean moisture transports as typical for the occurrence of heavy precipitation in the WMed. This study investigates the moisture origin for HPEs occurring during the extended winter season (September to February) as well as on the HPEs occurring in the eastern Mediterranean, where the wet season is during the winter months. There, heavy precipitation is dominated by the moisture supply of the Arabian Sea.

Other two factors very relevant for precipitating convection in the WMed are the high SST of the Mediterranean in autumn and the orography. The Mediterranean Sea, has a extension of about 2.5 millions km^2, containing 4.6 millions km^3 of water. By September it remains very warm with temperatures close to summer values (ca. $25°C$) (Pastor et al., 2017). Past studies, have stated that a progressive warming of $1.5°C$ has occurred for the whole Sea in the last

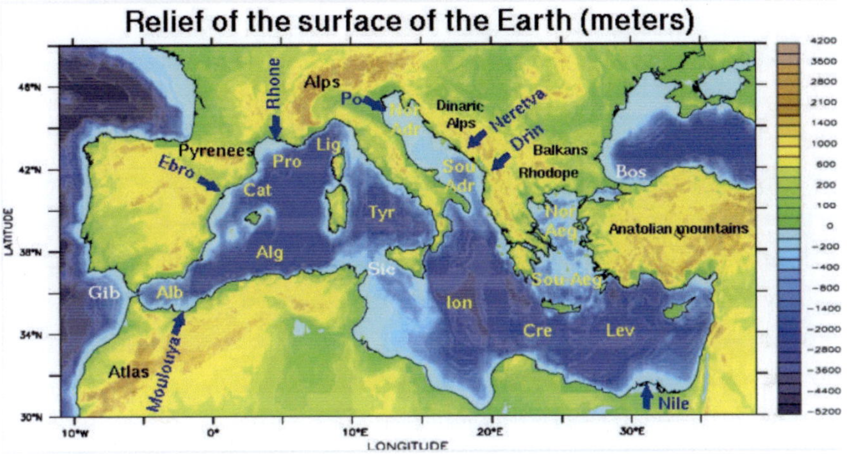

Figure 3.4: Relief of the Mediterranean basin (colour scale) and main geographical features. Obtained from Drobinski and Ducrocq (2007). The HyMeX White Book. Retrieved from *https://www.hymex.org/*. Last accessed 21-Nov-2018.

30 years. This is crucial for the heat and moisture release to the atmosphere through evapotranspiration which is larger than the fresh water inflow for the Sea. Indeed, the Mediterranean Sea is a concentration basin which means that it is characterized by a negative water budget (Lionello, 2012) due to a larger water loss through evaporation (E) than gain via other sources, such as precipitation (P). In addition to P, fresh water input comes from the runoff (R) of rivers surrounding the basin. Another two points of water exchange are the Black Sea, through the Aegean Sea (B) and the Atlantic Ocean through the Strait of Gibraltar (G). At the latter, relatively cool and saltier deep water flows out of the Mediterranean while in the upper sea-layers warmer and fresher water enters the Mediterranean basin. The mass (M) balance equation for the Mediterranean can be expressed as follows (Lionello, 2012):

$$\frac{dM}{dt} = G + B + R - (E - P) \tag{3.2}$$

Steep orographic features circle the Mediterranean Sea, from east to west along the northern basin and to the south over the mountains of Morocco Tunisia and Algeria (see Fig. 3.4). Particularly high are the Alps (this is the highest ridge with a maximum height of 4800m), and the Pyrenees systems, which are areas of heavy precipitation due to orographic forcing. Other mountain systems, where orographic precipitation is considerable are the Italian and the western Balkan ridges. Furthermore, about 200 Islands are distributed throughout the Sea, which, in some cases have a complex orography, such is the case for instance of the islands of Corsica and Sardinia. Hence is orography an important actor to DMC in the Mediterranean, through the aforementioned orographic precipitation mechanism, but also since it can act as a boundary to the mesoscale circulations redirecting the low level flows. Moreover, complex orography can also bring secondary circulations stemming from density heterogeneities and contrasts between soil type, land and sea surface or between elevated and low terrain. Such is the case, for example of the diurnal valley or the sea breezes.

Initiation, dynamics and shape of precipitating convection in the WMed

The presented characteristics of the Mediterranean basin, are factors conducive to heavy precipitation (Khodayar et al., 2016a; Ducrocq et al., 2016; Davolio et al., 2016). To summarize, these factors are, a favourable large-scale situation (favouring upper-level lifting, induced low-level circulations, or the build-up of cold pools, acting as mesoscale boundaries), wind circulations bringing sufficient moisture flux and eventually low-level convergence, high CAPE and a dynamic lifting mechanism (in the case of a strong inhibition), all of which may occur under slow-changing synoptic scale conditions extending the life cycle of DMC. The differences in total precipitation, and the evolution of each event, lie upon the distinctive characteristics of these factors and small variations in their timing or location.

These distinctive characteristics are the departure point for Davolio et al. (2016) in the assessment of HPEs affecting North-eastern Italy (NEI), originated by a strong south-easterly low-level inflow at the Adriatic Sea (Sirocco wind). Under two, apparently, similar synoptic and wind field situations, several autumn HPEs evolved in two very different manners. For the first group of three events, convection triggered at the NEI plains upstream of the Alpine ridge. On the contrary, for the second group (four events), the low-level flow was able to reach the mountain slopes being forced to lift and originating there orographic precipitation. The authors concluded that for the first group of events it was impossible for the sea south-easterly inflow to reach the Alpine area due to a blocking situation of cold air over the NEI plains due to larger evaporative cooling. On the contrary, the second group, showed precipitation occurring at the mountains given the cold pool at the NEI plains was less intense and the flow-over conditions developed, aiding the moist inflow to surpass the blocking and reach the Alpine foothills.

Another crucial aspect of the dynamics of DMC is the maintenance of the organized systems. Systems precipitating for very long times are associated with stationary situations, e.g. blocking, that can lead to heavy precipitation falling over the same location for several hours. Lee et al. (2016), addressed this issue focusing on two back-building Mesoscale Convective Systems (MCSs) growing over the Thyrrenean Sea on the 15 and 16 Oct 2012. These systems were very long-lasted (14h) since persistent moist inflow and blocking from the northerly cold front induced recursive triggering of new cells at almost the same location. This mechanism is known as back-building. Triggering of the first MCS started due to sufficient low-level convergence at the Thyrrenan sea, stemming from the confluence of a south westerly very moist flow, a southerly flow and a cold front descending from higher-latitudes. At this moment, high CAPE values ($> 1500\, J\, kg^{-1}$) and low-level moisture ($\approx 12\, g\, kg^{-1}$) were present, as well as a higher-level moisture plume originating over tropical Africa feeding the system. A second MCS triggered over Algeria, which showed very similar mechanisms, except for the fact hat it was less station-

ary. The persistent situation of the slow-moving northerly front, and of the southerly and southwesterly moist inflows together with the higher level moisture inflow, prevented the MCS from disappearing. The long-lasted precipitation terminated when a strong Mistral started at the end of 15-Oct.
Lastly, the influence of orography in initiating and maintaining convection during two HPEs occurring over Corsica (22-Oct-2012) and over eastern Spain (28 to 29-Sep-2012) (Ducrocq et al., 2016), is assessed. During the first HPE, the role of the complex orography of Corsica was crucial since it deflected the arriving northwesterly winds into two branches surrounding the mountain ridges. This resulted in lee-side convergence at the southern tip of the island continuously triggering new cells at the same location which only ceased when the eastern and western branches were disrupted. Regarding the HPE over eastern Spain, orography also played a decisive role in triggering convection. In this case the MCSs affecting the eastern Spanish coast were reinforced by lifting at the boundary of the eastern Spanish plains and the Iberian plateau, as demonstrated by Röhner et al. (2016) through modelling sensitivity experiments erasing the elevated terrain of the Iberian Plateau.

3.2 The HyMeX International Program and the SOP1 Field Campaign

HyMeX is an international research initiative aiming at a better quantification and understanding of the hydrological cycle and related processes in the Mediterranean (Drobinski et al., 2007). Special emphasis is put on high-impact weather events and their effects on the ecosystems and human activities of the Mediterranean regions. The HyMeX experiment has two global objectives

1 - Producing a new long-term and highly temporally and spatially resolved data set over the Mediterranean basin

2 - Developing methodologies and models in order to contribute to basic needs of weather prediction, regional climate studies, climate impact and environmental research

The HyMeX major experiment gathers scientists of more than ten European and African countries, offering an excellent scientific network for discussion and exchange. It is organized into five work packages, the water budget of the Mediterranean basin, the continental hydrological cycle and related water resources, heavy precipitation and flash-flooding, intense air-sea exchanges and coastal dynamics. HyMeX is thus an excellent science hub providing knowledge and novel observations to scientists of the community.

Overview on HyMeX-SOP1

Within the HyMeX experiment, a Selective Observation Period (SOP) was carried out with focus put on heavy precipitation, and gust events. This major field campaign took place between the 05-Sep and 06-Nov of 2012 in the WMed area and was operated by more than 80 institutions within HyMeX.
Eight target areas were defined to carry out intensive observation campaigns. Given the emphasis put on heavy precipitation these were selected given their tendency to be affected by this weather phenomenon. They are, therefore, areas prone to moist Mediterranean inflows or close to convergence areas and surrounded by complex orography. Figure 3.5 shows the eight selected areas, the Balearic Islands (BA), Catalonia (CA), Valencia (VA), Cevennes-Vivarais (CV), Corsica (CO) in France, Central Italy (CI), Liguria-Tuscany (LT) and NEI (Ducrocq et al., 2014). During the SOP1, maximum daily precipitation reaching 150 mm occurred over almost all target areas. Regions such as LT, or the Pyrenees reached 200mm and at CI, even one rain gauge collected more than 300mm in 24h. Monthly precipitation during Sep-2012 and Oct-2012 were well above the precipitation climatology (Ducrocq et al., 2014).
In addition to the fix instruments deployed during the SOP1 and the operational stations of collaborating weather institutions, supplementary measuring activ-

ities were carried out during selected dates, i.e. during Intensive Observation Periods (IOPs). During the IOPs, additional measurements such as Boundary Layer Pressurized Balloons (BLPBs) observations or aircraft flights, were carried out to gain a deeper insight of the processes forming heavy precipitation. Out of these 19 IOPs, 12 were dedicated to study events of heavy precipitation (HPEs) occurring between Sep and Nov of 2012. Figure 3.6 shows the temporal evolution of daily precipitation as obtained by the Climate Prediction Center Morphing Technique (CMORPH) product in black (see chapter 5 for a description of the data set), averaged over the whole WMed. Fig. 3.6 also shows the time spans of the 12 IOPs and the target areas selected for intensive measurements during the SOP1 campaign. Additionally, the dominating large-scale situation, when these took place is shown at the top of the axes. This information is obtained from Ducrocq et al. (2014). The first events, between 5 to 22-September, took place with an Atlantic ridge regime dominating the Wmed. This regime shifted on the 23 to 30-Sep to a negative phase of

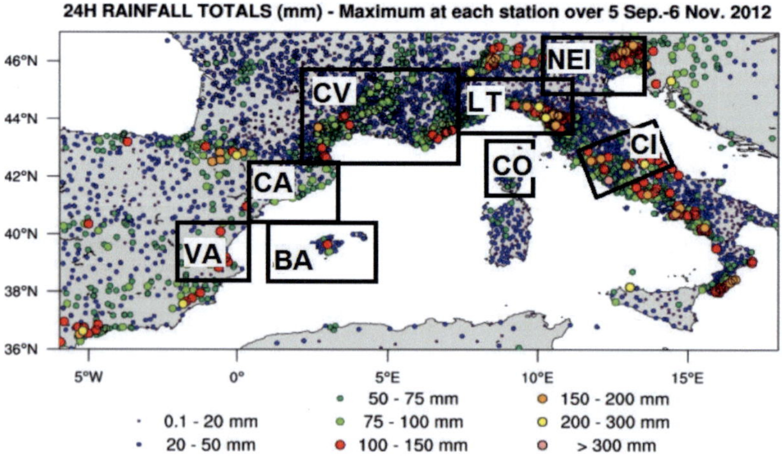

Figure 3.5: Spatial distribution of precipitation during the SOP1 campaign (05-Sep to 06-Nov-2012) and the eight HyMeX investigation areas. Obtained from Ducrocq et al. (2014).

the North Atlantic Oscillation (NAO), providing favourable large-scale lifting. Between 16-Sep to 03-Oct Hurricane Nadine meandered around the Azores region, to switch to a positive phase of the NAO between 16-Oct to 05-Nov (see Fig. 3.6).

The HyMeX-SOP1 campaign is a benchmark for DMC studies in the Mediterranean given the vast number of instruments deployed during an exceptionally long period, including state-of-art technologies, among which several were exclusively dedicated to the retrieval of water vapour in the atmosphere. Among

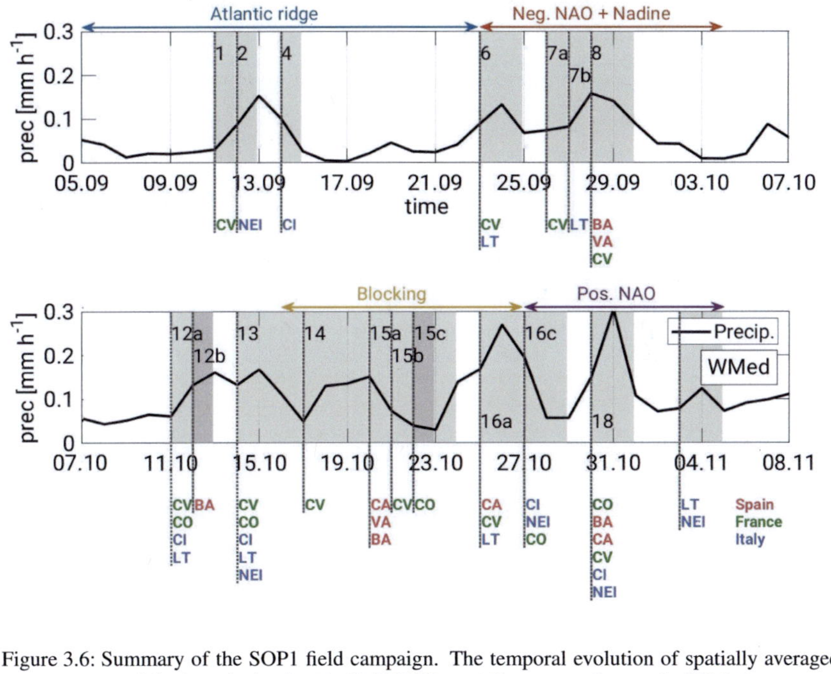

Figure 3.6: Summary of the SOP1 field campaign. The temporal evolution of spatially averaged precipitation obtained with CMORPH spatially averaged over the WMed are represented by the thick black line. The dates of the HyMeX IOPs (gray shading) and the corresponding affected target areas (red, green, blue colours), are likewise presented. The dominating large-scale situation as addressed by Ducrocq et al. (2014) is presented at the top of the axes.

others, during the SOP1 campaign the following instruments were deployed: more than five S-band and C-band Doppler radars, rain gauges, wind profilers, fix and mobile radiosonde stations, Lightning Mapping Arrays (LMAs), GPS-receivers, disdrometers, MicroRain Radars, instrumented aircrafts and pressurized balloons, the *Lidar pour l'étude des intéractions Aérosols Nuages Dynamique Rayonnement et du cycle de l'Eau* (LEANDRE 2), aboard the aircraft ATR-42, advanced microphysical in-situ probes and a 95-GHz Doppler cloud radar named *Radar Aéroporté et Sol de Télédétection des propriétés nuAgeuses* (RASTA) aboard the F20 research aircraft.

4 The COSMO Model and the Nudging Scheme

Numerical weather and climate prediction modelling is a key tool in the prediction and prevention of weather hazards associated with heavy precipitation and flash flooding. In this thesis numerical prediction is used to characterize possible biases of a well-established model as well as to advance the understanding on the moist processes related to Deep Moist Convection (DMC). It is used as a tool to perform experiments inducing corrections in the water vapour fields by means of a Data Assimilation method.

The used prediction system is the Consortium for Small-scale Modelling (COSMO) model (Baldauf et al., 2011), and both its Numerical Weather Prediction (NWP) mode (COSMO) and Climate mode (COSMO-CLM) are used. COSMO has been selected given it is a well-established Limited Area Model (LAM) used operationally in several European countries with more than 12 partners institutions and collaborators. Two of its main advantages, exploited in this thesis, are, on the one hand, the availability of a Data Assimilation (DA) scheme, called the Nudging scheme, especially suited for assimilation of asynoptic data which has shown especially good scores in correcting humidity fields (Schraff et al., 2016). The second advantage is its usability on several grid sizes, favouring the description of atmospheric phenomena spanning several scales, from the Macro-β to the Micro-α.

In this chapter the modelling technical aspects relevant to this thesis are introduced with focus put on the model description, the representation of convection and the assimilation of humidity observations.

4.1 The COSMO Model

The COSMO model is a non-hydrostatic Numerical Weather Prediction (NWP) model based on the primitive thermo-hydrodynamical equations (see Fig. 4.1) for a compressible flow in a moist atmosphere (Baldauf et al., 2011). The atmosphere in COSMO is described mathematically as a multicomponent continuum constituted by dry air, water vapour, liquid water and water in solid state in an idealized mixture (Doms et al., 2011). The atmospheric system is subject to gravity and Coriolis Forces, where the heat, momentum and mass processes are subject to the conservation laws. Several assumptions are used to derive the basic prognostic equations, always respecting compressibility. These simplifications include, the neglecting of the molecular fluxes (except for the diffusion of liquid and solid water), assuming the specific heat of moist air to be that of dry air, neglecting constituents resulting from diffusion fluxes and phase transitions and finally neglecting temperature changes due to buoyant heat and moisture fluxes. After these simplifications the following set of equations describes the non-hydrostatic compressible mean flow:

$$\rho(d\boldsymbol{v}/dt) = -\nabla p + \rho \boldsymbol{g} - 2\Omega \times (\rho \boldsymbol{v}) + \nabla \cdot (\boldsymbol{T}) \tag{4.1a}$$

$$\frac{dp}{dt} = -(c_{pd}/c_{vd})p\nabla \cdot \boldsymbol{v} + (c_{p,d}/c_{v,d} - 1)Q_h \tag{4.1b}$$

$$\rho c_{pd}\frac{dT}{dt} = \frac{dp}{dt} + Q_h \tag{4.1c}$$

$$\rho \frac{dq^v}{dt} = -\nabla \cdot \boldsymbol{F}^v - (I^l + I^f) \tag{4.1d}$$

$$\rho \frac{dq^{l,f}}{dt} = -\nabla \cdot (\boldsymbol{P}^{l,f} + \boldsymbol{F}^{l,f}) + I^{l,f} \tag{4.1e}$$

$$\rho = pR_d(1 + (R_v/R_d - 1)q^v - q^l - q^f)T^{-1} \tag{4.1f}$$

Where a) is the momentum equation b) is the pressure tendency, c) is the budget equation, d) and e) are the mass conservation equations for vapour and condensate, respectively and f) is the equation of state. With $\boldsymbol{v}$, ρ, p, T,$\boldsymbol{g}$, q^x, c_{xd}, Ω as the winds, density, pressure, temperature, gravity acceleration, mass of vapour and condensate, specific heats of dry air and rotation rate of the Earth. And Q_h, $\boldsymbol{T}$, I^x, F^x, P^x and R_x as the diabatic heat production, the Reynolds stress tensor, the sources/sinks of a constituent, the turbulent fluxes, the precipitation fluxes and the gas constant, respectively.

The COSMO model takes into account the unresolved processes by including physical parameterizations for subgrid-scale turbulence, the surface layer, clouds and precipitation, convection, radiation and soil processes. Regarding the model grid, it is an Arkawa-C type and can be rotated to avoid problems emerging from meridian convergence near the poles. The vertical distribution of model levels uses the Gal-Chen terrain following formulation. It is based on a coordinate system transformation performed in two steps. First, the terrain following coordinate for height z, is transformed into a user-defined height coordinate $\overline{\zeta}$, where the relation between z and $\overline{\zeta}$ is the monotonic function z=f(λ,ψ,$\overline{\zeta}$). Second $\overline{\zeta}$ is mapped to the computational coordinate ζ by the transformation $\overline{\zeta} = m(\zeta)$. Where m is chosen a monotonic discrete function that renders the vertical coordinate a rectangular grid, labelled by integers (see Fig. 4.1). The time-integration scheme is a 2 time-level Runge-Kutta with time-split treatment of acoustic and gravity waves and the model prognostic variables can be extended to cloud ice content and specific water content of rain, snow and graupel.

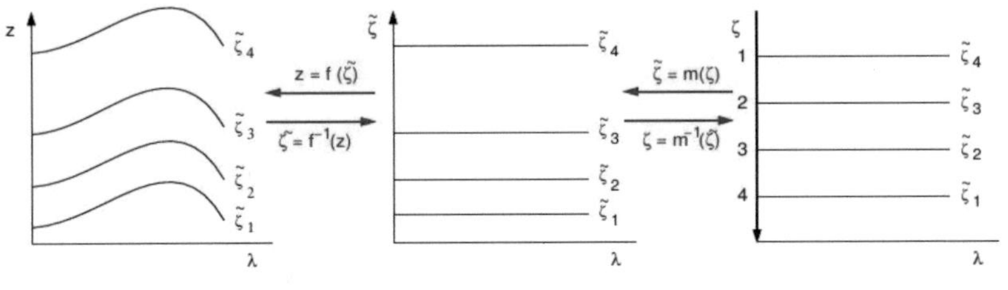

Figure 4.1: Mapping of an irregular curvilinear grid associated with the terrain following coordinate onto a rectangular equidistant grid.

As a LAM, COSMO needs forcing data from coarser resolution models (typically global) to act as initial and boundary conditions to the atmospheric differential equations. One of COSMOs main advantages is its versatility in reading data from several atmospheric global models (GME, Integrated Forecast System; IFS from the European Centre for Medium-Range Weather Forecasts ECMWF as well as from coarser resolution simulations performed with COSMO itself). At the lateral and upper boundaries it considers a sponge layer with Rayleigh damping at the upper levels to guarantee a smooth coupling of the coarser model data with COSMO. This allows the one-way nesting of coarser model data which is a technique commonly used for dynamic downscalling (Chan et al., 2012; Coppola et al., 2018). As shown in Figure 4.2 very high model resolutions (≈ 1 *km*) are reached by nesting the finer grids into the coarser resolution simulations.

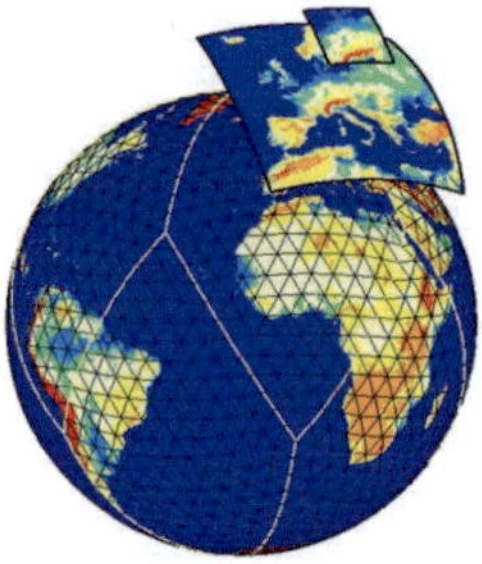

Figure 4.2: Concept of dynamical downscalling. Coarser resolution models issue the forcing data for finer grid simulations. Obtained from textithttps://www.dwd.de.

At the bottom of the mesh, the interaction between the soil and the atmosphere, through fluxes and heat and moisture exchanges, is controlled by the TERRA-ML (Doms et al., 2011) Soil-Vegetation-Atmosphere-Transfer (SVAT) scheme. This scheme uses soil and vegetation parameters, such as Leaf Area Index (LAI), plant cover, root depth, roughness-length and soil type read from external data sets. In the case of vegetation, these are provided by the Global Land Cover 2000 (GLC2000) data set (Latifovic et al., 2004), the orography is given by the Global Land One-km Base Elevation Project (GLOBE) data set (Hastings et al., 1999), finally the soil type is provided by the FAO Digital Soil Map of the World (DSMW) (FAO, 1995).

At the lower meso-γ limit between 3 km and 1 km deep convection and/or shallow convection are explicitly resolved. Simulations in convection-permitting grids have helped overcome well-known biases of large-scale simulations: misrepresentation of the diurnal cycle of precipitation, underestimation of dry-days and of precipitation intensities and overestimation of low-precipitation event frequency (Prein et al., 2015; Fosser et al., 2016). The processes occurring at small-scales are resolved and specially, the interaction of the atmosphere with complex orography and soil heterogeneities, both of which are important factors for convection triggering (Chan et al., 2012). Added value has also been shown in the representation of humidity and temperature vertical profiles (Fosser et al., 2014), besides a more realistic representation of precipitation patterns i.e. larger precipitation maxima and smaller structures. On-going research in the field focuses on the adaptation of the parameterization schemes (turbulence, radiation, clouds) to these finer scales favouring a seamless simulation approach, given they were conceived for coarser model resolutions.

Given some atmospheric processes occur at scales finer than that of the grid length, the model is unable to explicitly resolve them and needs physical parameterizations. The parameterization schemes included in COSMO are: clouds, precipitation, atmosphere-soil interactions, radiation, turbulence and convection. The latter will be explained in more detail further in this section. The clouds and precipitation scheme is a two-moment water continuity model which, in the configuration used in this thesis, uses dependent variables for cloud water, precipitable water, cloud ice, snow and graupel (for very high-resolution simulations). The inclusion of these hydrometeor categories allows the representation of precipitation in water, mixed phase, graupel and ice phases. The radiative transfer scheme is the Ritter and Geleyn (1992) scheme, also called δ-two-stream solution of the radiative transfer equation for a plane-parallel horizontally homogeneous atmosphere. It uses 8 spectral intervals and interacts with clouds, water vapour, ozone, CO_2 and aerosols as optical constituents. Turbulence is parameterized following the Mellor and Yamada Scheme (1974) based on a second-order closure. The scheme is based on the Turbulent Kinetic Energy (TKE) closure and has both a 1D application i.e. with homogeneous horizontal surfaces and neglecting turbulent diffusion in the horizontal direction and a 3D application. The 3D TKE closure scheme is to be applied on simulations with horizontal resolutions ($< 1\ km$), at those scales the boundary layer approximation is no longer valid and turbulent diffusion in the horizontal direction is also needed. The application of a 3D TKE closure has shown to enhance mixing and microphysical processes bringing more precipitation in case studies with a better agreement to observations (Verrelle et al., 2014).

Parameterized and Explicit Convection

Atmospheric convection in various forms ranges weather scales from the meso-γ (spatial scales $\approx 2\ km$, temporal scales <1h), for example as isolated thermals, to the organized meso-α convective systems (spatial scales $\approx 10^2\ km$, temporal scales of days). Examples of these are squall lines or Mesoscale Convective Systems (MCSs). Given the spatial scale of the convective updrafts is of only a few kilometers, weather and climate prediction models cannot explicitly resolve them in the case of having grid lengths larger than 3km. The parameterization scheme implemented in COSMO is a low-level control, mass-flux type (Tiedtke, 1989) which calculates the mass heat, moisture and momentum transfers by using a bulk cloud model. These so-called low-level control schemes focus on how the low-level air parcels could potentially overcome CIN to release the existing Convective Available Potential Energy (CAPE). In particular, the Tiedtke Scheme is very dependent on the moisture supply from large scale convergence and from boundary layer moisture and momentum transports, necessary to trigger the updraughts (Doms et al., 2011). When the grid sizes are finer than a few kilometers ($\approx\ 3\ km$), deep convection is explicitly resolved.

Hence a parameterization for this phenomenon is no longer needed and can be switched off. Still, shallow convection still needs to be parameterized. Only when sub-kilometer grid lengths are reached can the models represent shallow convection.

The Numerical Weather and Climate Prediction modes

Variables such as ozone and aerosol concentration, or canopy variables can be approximated as constant in simulations of a few days (event-scale). However, when longer periods are considered (seasonal, yearly), these have to be adapted to account for their yearly variation. COSMO, has a Numerical Weather Prediction (NWP) version as well as a climate version where these remarks are considered. The Climate Limited area Modelling Community (COSMO-CLM), a partner initiative of the COSMO consortium, develops and maintains a version of the COSMO source code which allows for longer runs

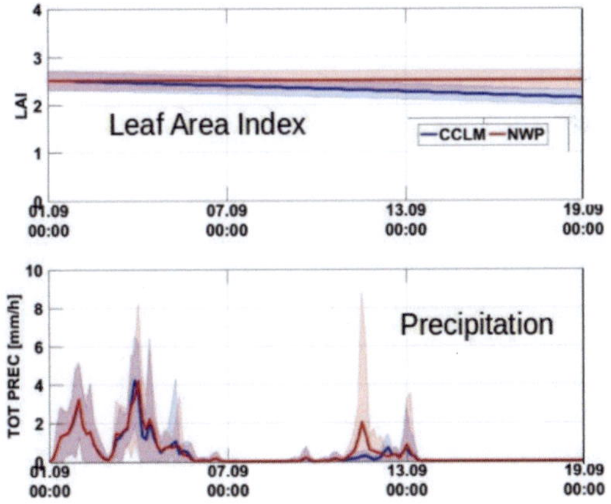

Figure 4.3: Comparison between COSMO in climate against COSMO in NWP mode. LAI is treated as a constant parameter in the NWP configuration, whereas in the climate version it is updated. The spatially averaged temporal evolution of LAI and precipitation is shown.

to be carried out (beyond the medium range). This is the so-called COSMO in Climate Mode (CCLM). In this configuration, vegetation variables (LAI, root depth and plant cover), in addition to other slowly-varying variables such as the aforementioned ozone species and temperature of the surface and of snow are updated from the external parameters files.
Figure 4.3 shows the temporal evolution of one of the slow-changing variables (LAI) and of precipitation over a study region in Europe. It can be seen how after ca. 6 days of simulation the COSMO-CLM LAI starts to diverge from the NWP. This has a large impact on simulated precipitation beyond day 10.
The numeric and dynamical cores of the weather and the climate version are exactly the same. These two versions only differ in the adaptation of the slowly varying variables.

The water cycle in COSMO

In the atmosphere, moisture is represented in the form of water vapour, liquid water and water in solid state. Several species of water in solid state are considered in COSMO such as cloud droplets, raindrops, ice crystals, rimed aggregates graupel or hail. The basic set of equations (See Eq. 4.1) offers a prognostic solution to the dynamics of water (in all its states), by means of the mass conservation law.
The exchange of moisture with the soil and water surfaces is parameterized in COSMO. External parameters and forcing data provide the initial and boundary conditions of temperature and moisture at the soil and the sea to parameterize the surface fluxes.
The exchange of moisture with the surface of the model is carried out by means of the Multi-Layer Soil and Vegetation Model (TERRA-ML). This is the scheme responsible for coupling the atmosphere with the sea and the soil. TERRA-ML uses a multi-layer structure that models the coupling between the atmosphere and the surface using a stability and roughness-length parameters

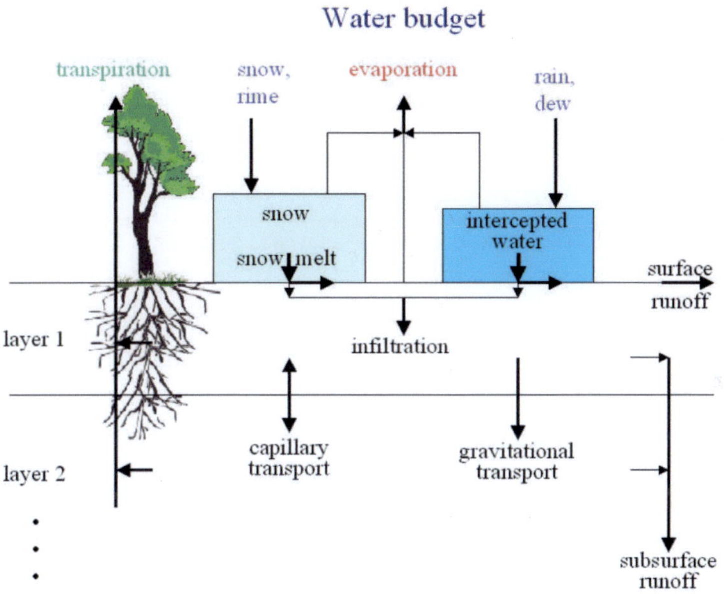

Figure 4.4: Concept of the hydrological processes considered in TERRA-ML for moisture gains and losses. Obtained from Doms et al. (2011).

dependent on the flux formulation. The soil parameterization has eight defined soil types (sand, sandy loam, loam, loamy clay, clay, ice, rock and peat).

TERRA-ML considers two processes as sources of moisture to the soil, precipitation and formation of dew and rime, and three processes as sinks. The sinks are evaporation, transpiration and water runoff. The processes are summarized in Fig. 4.4. The exchange of moisture and heat to the atmosphere is performed by the calculation of the drag fluxes, and towards the ground by solving the heat condition and Richards equation. For water infiltration to deeper levels in the soil, the Richards equation considers 6 active layers, where water transport vertically is computed for each of the layers.

The amount of sensible/latent heat flux absorbed or released at the surface-atmosphere boundary depends on the gradient of humidity and temperature between the bottom atmospheric model layer and the soil surface. The COSMO model parameterizes these exchanges in the surface flux of sensible heat (H_{sfc}) and the surface flux of water vapour (F^q_{sfc}) as proportional to the temperature and humidity differences between the ground and the atmosphere.

$$H_{sfc} = -\rho C^d_h |v_h| \left(\theta \pi_{sfc} - T_{sfc}\right) \quad (4.2a)$$

$$F^q_{sfc} = -\rho C^d_q |v_h| \left(q - q_{sfc}\right) \quad (4.2b)$$

Where ρ is the air density, C^d_h and C^d_v are the bulk-aerodynamical transfer coefficients for turbulent heat exchange, θ and π_{sfc} are the potential temperature at the lowest grid level and the scale pressure at the ground, respectively, T_{sfc} is the ground temperature and q and q_{sfc} are the specific humidity at the lowest grid level and the surface ground. This parameterization moistens or heats the soil or the lower atmosphere when they are either drier or cooler, as in a diffusion problem. Over the sea, the formulation is analogous, save that the drag coefficient for the moisture fluxes is different to account for water instead of soil type.

4.2 Data Assimilation and the Nudging Scheme

Numerical weather and climate prediction is an initial value problem (Franke, 1988) where the quality of the predictions depends crucially on the accuracy of the issued initial and boundary data. Observations are used, in this regard to bring weather and climate forecasts closer to reality. In limited area operational modelling, usually, the simulations contain observation information at two stages. The first stage, is the forcing of the regional model using global data which already contain observations. At the second stage, the simulations are corrected by the input of observations in the so-called Data Assimilation (DA) cycles. There are, at present, several DA methods (Nudging, 3DVar, 4DVar, Optimal Interpolation), each of which shows its advantages and drawbacks. The core of many of these methods is the analysis equation

$$\boldsymbol{x}_a - \boldsymbol{x}_b = \boldsymbol{W}[\boldsymbol{y}_O - H(\boldsymbol{x}_b)] \tag{4.3a}$$

$$\boldsymbol{W} = \boldsymbol{B}\boldsymbol{H}^T(\boldsymbol{R} + \boldsymbol{H}\boldsymbol{B}\boldsymbol{H}^T)^{-1} \tag{4.3b}$$

Where $\boldsymbol{x}_a$ is the analysis state of the prognostic model variable $\boldsymbol{x}$, $\boldsymbol{x}_b$ is the background, $\boldsymbol{W}$ is the gain matrix which depends on the covariance of the model background ($\boldsymbol{B}$), of the observations error ($\boldsymbol{R}$) and the lineal forward operator of the observations ($\boldsymbol{H}$), and finally $\boldsymbol{y}_O$ are the observations. The special features of the different data assimilation methods arise from how the gain matrix is obtained. In the Optimum Interpolation method, for example, $\boldsymbol{W}$ is obtained by minimizing the covariance matrix of the analysis error. This allows for consideration of the spatial correlations between the model variables. However, it includes stringent simplifications such as representing the background correlation as homogeneous and isotropic. In the variational schemes, the analysis equation is reinterpreted in the form of a cost function, consisting of two terms, one for the analysis increments in the model space and the other in the observation space, which is minimized. This approach allows avoiding simplifications of the optimum interpolation method, and are better applicable to satellite data.

In the 3DVar scheme, the main drawback is that it does not propagate the uncertainty during the assimilation time window, neither it accounts for the temporal evolution within the assimilation time window. The 4DVar scheme depends on time and therefore the temporal dependence is considered but it does not transport the uncertainty. Another inconvenient of the 4DVar scheme is that it can be computationally very costly. Indeed within each assimilation time window the model state is propagated forward and backward by means of a linear operator for the model. A workaround to reduce computational costs is the Ensemble Kalman Filter (EnKF) which uses an ensemble of simulations to calculate the background covariance matrix. Moreover, the background error covariances and the state of the model are propagated with time using the model equations

(Hunt et al., 2007). This allows to propagate the model uncertainty, without needing to linearize the atmospheric model and also, with a sufficient number of ensemble members, it shows analogue results to the 4DVar.
The DA method implemented in COSMO and employed and assessed throughout this thesis is the Nudging scheme as described by Schraff and Hess (2012). This scheme was selected given it was used operationally at the German Weather Service (DWD) until April 2017 for COSMO simulations, before the implementation of the KENDA system. It is a well-established scheme and the extensive know-how at the DWD makes its tuning most suited for the moisture sensitivity experiments. Recent comparative studies (Schraff et al., 2016) between the Nudging scheme and the, now, operational system at the German Weather Service (DWD), the Kilometre-scale Ensemble Data Assimilation (KENDA) system (using a Local Ensemble Transform Kalman filter; LETKF) have shown that the nudging scheme is hard to outperform for upper-level humidity. The election of the scheme is justified since, the primary question in the experimental set-ups presented in this thesis, is the achievement of the most realistic humidity fields, by means of data assimilation, to thereafter study the impact on the evolution of convection.

The Nudging Scheme

The main difference of the Nudging Scheme with respect to the schemes presented in the last subsection is that it is an online DA method. The nudging scheme does not need to compute the background covariances of the model in order to know the state of the atmospheric system. Instead, it corrects the fields of the assimilated variables at correct time steps. The main feature of this system is its simplicity, since it avoids large departures of the model state from reality by applying stringent bias and quality checks. The quality and bias checks impede the assimilation of observations suspicious of being erroneous.
The Nudging scheme, also called Newtonian relaxation, consists of relaxing the model's prognostic variables towards prescribed values within a predeter-

mined time window (Schraff et al., 2013). This is done by adding an extra term to the numerical solutions, which depends on a coefficient, G_ψ, the model-observations difference, $\left[\psi_k^{obs} - \psi_k^{mod}\right]$ (the observation increments), and on weight functions balancing the impact of the assimilation depending on the location, time and quality of the input observation, W_k.

$$\frac{\partial}{\partial t}\psi(\boldsymbol{x},t) = F(\boldsymbol{\psi},\boldsymbol{x},t) + G_\psi \cdot \sum_{k_{obs}} W_k(\boldsymbol{x},t) \cdot \left[\psi_k^{obs} - \psi(\boldsymbol{x}_k,t)\right], \tag{4.4}$$

$\psi(\boldsymbol{x},t)$ represents a model prognostic variable, $\boldsymbol{x}$ and t the space and time coordinates, and k is a running index for the observations. The Nudging coefficient has units of frequency and its used to control the extent of the impact of the second term in the right hand side of the equation. The weight function is computed as follows, for multiple observation types

$$W_k = \frac{w_k}{\sum_j w_j} \cdot w_k, \quad with \quad w_k = w_t \cdot w_{xy} \cdot w_z \cdot \varepsilon_k \tag{4.5}$$

Where the weight for a single observation (w_k) depends on the quality and representativeness weights for each observation type (ε_k), the spatial distance (w_{xy}, w_z), and temporal difference (w_t), between the observation and the target grid point. The observation increments are computed during the assimilation time window and are ingested into the solutions of the dynamic and numeric routines. Consequently, step by step the forecast is "dragged" towards the observed values as shown in Fig. 4.5.

The quality weights are different for each observation type. In the particular case of Zenith Total Delay (ZTD) (or equivalently Integrated Water vapour (IWV)), ε depends on the layer thickness and temperature. Provided that thin or very cold model layers contribute only little to IWV, they should not be influenced by the IWV observations as much as thick and warm layers. Hence for this measurement the quality weights (ε) are proportional to saturation specific humidity ($q_{v_{sat}}$) and layer thickness (Δz). For each model level (k) they are defined as

$$\varepsilon_k = \frac{\Delta z_k \cdot q_{v_{sat_k}}}{max_{k'}\left(\Delta z_{k'} \cdot q_{v_{sat_{k'}}}\right)} \leq 1, \tag{4.6}$$

Concerning the lateral spreading weight, it is a second-order autoregressive function of the distance Δr between the observation location and the target point (Schraff and Hess, 2012).

$$w_{xy} = (1 + \Delta r/s) \cdot e^{-\Delta r/s} \tag{4.7}$$

Where w_{xy} is the weight for horizontal spreading, Δr is the distance between the observation location and the target grid point, and s is the scale separation for the nudged variable. For small separations, $(\Delta r/s)$ falls off a little faster with distance than a Gaussian until a certain radius of influence is reached. This is set, for humidity to $2 \cdot s$ and for temperature to $3.5 \cdot s$. Thus, at a level of 1000 hPa the radius of influence is cut at about 116 km for humidity and at 291 km for temperature. Higher in the troposphere, at 500 hPa, is set to 166 km for humidity and 291 km for temperature.

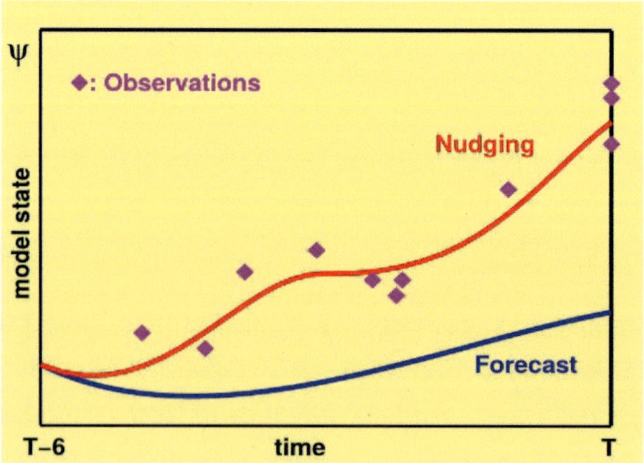

Figure 4.5: Conceptual illustration of the effect of the Nudging Scheme (Schraff and Hess, 2012).

In the Data Assimilation process, the reports are read from observation input files, usually in NetCDF format, three to four hours ahead of the simulation time. To present, the Nudging scheme admits data from radiosondes, aircraft observations, wind-profilers, surface-level observations, upper-air wind from Doppler radar, Radio Acoustic Sounding Systems (RASS), Sonic Detection and Ranging (SODAR), Global Positioning System-derived IWV (GPS-IWV), scatterometer 10-m wind data and atmospheric motion vectors (Schraff and Hess, 2012). Once allocated the data, the assimilation routines that compute and add the analysis increments are called periodically, depending on the assimilation time window. Before their acceptance, the observations are assigned temporally and spatially to a point in the model space. The quality flags, the blacklists, and the user-defined priority observation types, sort and neglect measurements depending on their reported accuracy. Atmospheric profiles are built and put together in order to assimilate prognostic variables and finally, bias corrections, quality control, gross error check and redundancy tests are carried out to neglect suspicious measurements.

A crucial step is the assignment of observations to the model grid points at the corresponding simulation hour. Given that in general the input observations are not strictly simultaneous to the time step being computed at run time, the observations need to be spread in time. This is done in two different ways. For very frequent observations (less than an hour) the atmospheric evolution is considered linear and the information is interpolated in time to the model time step. For observations of coarser frequency, a time weight function is applied. This function is continuous, asymmetric and its e-folding decay should be slightly larger than the relaxation time of the assimilation (this is given by the inverse of the nudging coefficient) it should be of about 1 to 2 hours. Spatially, first, the measurements are assigned to the closest grid point in the horizontal dimension provided that the distance between the selected grid-point and the observation location is $< 0.5\Delta x$ in the horizontal and less than 40m in height with Δx the model horizontal resolution. This is acceptable since grid point models cannot correctly represent wave lengths of $2\Delta x$ or less.

Regarding the vertical spreading (Eq. 4.8), the weight function (w_z) spreads the information in the vertical direction, following a Gaussian, depending on the distance between the model and the observation height ($\Delta z = z - z_{obs}$), the correlation scale ($ln\ p_c$), the gas constant (R) and the virtual temperature at the uppermost increment level ($T_v^{z_{obs}}$). Before spreading the information in the vertical, the observation is assigned to a grid-point only if the height distance of the report surface and model surface lays within the range 400 to -100m. In other words, as long as the model surface is as much as 100m below the station height or 400m above.

$$w_z = exp\left(\frac{-\left[g/RT_v^{z_{obs}}\right]}{ln\ p_c}\right)^2, \tag{4.8}$$

Nudging atmospheric moisture observations

Special focus is put in this thesis in the nudging of column atmospheric moisture observations. Global Positioning System (GPS) technology provides accurate, fast and affordable measurements of the total column moisture in the zenith direction. As will be introduced in Chapter 5.3 GPS receivers measure the delay endured by the microwave emitted signal from the GPS satellite constellation. This quantity is called ZTD.

ZTDs can be converted to IWV measurements as described in Bevis et al. (2012), provided the temperature and the pressure at station height are known IWV and ZTD are two interchangeable quantities. In COSMO, pressure and temperature are interpolated or extrapolated to the station location from the model values at run time. Once the IWV value has been obtained and assigned to a grid point it needs to be converted to a model specific prognostic variable. In the case of IWV it is converted to atmospheric specific humidity. Since IWV is an integrated value a profile has to be built up. This is done by means of an iterative process where the simulated specific humidity profile at a selected grid point ($q_{v_{mod}}$) and the corresponding IWV (IWV_{mod}), values are scaled to the input IWV observation, (IWV_{obs}).

$$q_{v_{obs}} = q_{v_{mod}} \cdot \frac{IWV_{obs}}{IWV_{mod}}, \tag{4.9}$$

The process continues until the modified profile, $q_{v_{obs}}$, attains an IWV of $0.1\% IWV_{obs}$, or until the 20th iteration is reached. From this moment on, the assimilation of the profiles from soundings and IWV is the same

Quality and Bias Checks

Column atmospheric moisture observations converted into profiles and humidity data from radiosondes are assigned to a specific grid point. For both observation types analogue quality controls and checks for suspicious data rejection. Since humidity assimilation in the profile influences the vertical distribution of temperature, the superadiabatic lapse-rate is checked in order to find unrealistically high increasing rates. Subsequently, the observation values can be interpolated back to the model levels. This is done by applying a vertical scale adjustment (Woodage, 1985) a method, which in comparison to a simple interpolation is hydrostatically consistent. After assignment of the spatial and quality weights the nudging is performed. Some radiosonde-specific consistency and error checks include, rejection of reports where the dew point temperature is lower than 150°C or larger than 40°C and supersaturations of $RH > 120\%$. A bias correction for humidity is also applied for radiosonde data given most sounding systems seem to show a dry bias near saturation. In such case saturation is considered to occur at $RH = 96\%$ irrespective of temperature.
Together with the construction of the specific humidity profile, a major difficulty of ZTD (IWV) assimilation, is the agreement between the model and the observation height. About 50% of the water vapour is concentrated within the first 1.5 km (Munn et al., 2003). This implies that small differences in what is considered as the station height impact largely the modelled IWV (Bock et al., 2016). To minimize such deficiencies, COSMO applies a height difference threshold in which reports are rejected if this difference is outside the range

400 m to -100 m, as explained earlier in this chapter. The surface observations lying within this range, are subsequently interpolated or extrapolated to the model height. Additionally, gross error checks are performed for the IWV observations such as rejection of values which are too low (<2 mm). Besides, COSMO checks for redundant reports and dismisses them, also it prioritizing reports coming from good quality stations.

5 Methodology, Experiments and Observational Data sets

5.1 Methods, Definitions and Model Validation

The set of methodologies applied in this thesis are introduced in this chapter. They are conceived to address the research objective presented in Chap. 1. The methodology focuses on the study of the sensitivity of the represented convective processes and precipitation to changes in the spatio-temporal distribution of atmospheric moisture. These changes are induced by the nudging scheme, used to bring the humidity fields to a more realistic distribution in these humidity nudging experiments. Hence this Data Assimilation (DA) method is not used in the operational way, to evaluate the accuracy of the forecast ranges, but rather to gain an improved representation of humidity. Indeed the nudging is performed continuously and every ten minutes new observations are input. These are called hereafter column moisture nudging experiments. The different simulation experiments share a core approach, which can be summarized in the following step chain. Reference runs are produced to simulate a particular event or relevant period for heavy precipitation. These are validated against observations. A subsequent simulation (or series of subsequent simulations) is carried out nudging humidity data, aiming at assessing the impact in a particular aspect. Then, the modified runs are likewise validated against observations. Last, a comparison between the reference and the modified runs is carried quantitatively and quantitatively using statistical methods, specific metrics, process-representation, etc.

The assessment on the response of the model to the atmospheric moisture correction by the Nudging scheme is carried out for simulations on several model grid lengths. In particular the grid lengths used are of 7 km, 2.8 km and 500 m. These grid lengths are used given they are characteristic of model configurations using a deep convection parameterization scheme (7 km), a shallow convection parameterization scheme (2.8 km) only and the use of a 3D TKE closure for the turbulence scheme with no convection parameterization (500 m).

Refining the model grid makes possible the study of the Deep Moist Convection (DMC) representation using a multi-scale approach. This is of main importance given the broad range of scales where precipitating convection triggers and evolves ($\approx 10^0$ km to 10^2 km, $days$ to $sub-hourly$) and where atmospheric moisture transports take place. The whole HyMeX-SOP1 (1-Sep to 20-Nov, 2012) period is reproduced with the Consortium for Small-scale Modelling (COSMO) in its climate version (COSMO-CLM), as well as chosen Heavy Precipitation Events (HPEs) in the Numerical Weather Prediction (NWP) version of COSMO. An ingredients-based approach for studying convection (Doswell et al., 1996) is followed. In the following, a glossary of relevant definitions used in this thesis is presented, as well as a technical description of the simulations of the HyMeX-SOP1 period and those of a selected case study IOP6.

Definitions

- Heavy Precipitation

Events showing precipitation rates larger than 100 $mm\ d^{-1}$ or 4.2 $mm\ h^{-1}$ are considered HPEs, for the western Mediterranean region (WMed) and the period of study (autumn 2012). Past studies over this area and season have reported HPEs with values ranging from 120 $mm\ d^{-1}$ (Chazette et al., 2015a; Lee et al., 2017) up to 200 $mm\ d^{-1}$ (Amengual et al., 2015; Bouin et al., 2017). Additionally, a second definition for heavy precipitation is considered. This is the all-day percentile intensities as suggested by Schär et al. (2016) where the value of heavy precipitation will be given by the percentile-90, accordingly to previous studies (Allan and Soden, 2008; Scoccimarro et al., 2013; Fischer and Knutti, 2016).

- Degree of synoptic forcing

As introduced in Sec.2.2 the convective adjustment time-scale (τ; Molini et al. (2011) and Keil et al. (2013)) gives a good measure of how fast convection is able to remove existing Convective Available Potential Energy (CAPE) and whether this happens at time-scales of convective updrafts (a few hours) or in longer time scales. In this thesis two categories of the continuum scale of τ are defined. An event is considered strongly synoptically forced if τ<3h. An event is considered weakly synoptically forced if τ>12 h in the sense of having a large influence on the mesoscale factors creating CAPE. This is a continuum hence cases in between are also possible.

- Preconditions leading to Deep Moist Convection

Doswell et al. (1996) suggested an ingredients-based approach to determine the regions prone to heavy precipitation, useful for forecasters and prevention. Three elements are needed for DMC to develop, namely atmospheric instability, sufficient moisture amount and a triggering mechanism aiding to overcome the possible existing inhibitions. Past studies define preconditioning of heavy precipitation in varied ways. Hohenegger and Stevens (2013), for example, defined preconditioning of DMC as moistening of the lower free-troposphere by growing cumulus congestus evolving to deep precipitating cumulonimbi. Soderholm et al. (2017), in a case study of subtropical convection, pointed at sea-breezes, near-surface variations of moisture and diurnal heating as the elements preconditioning the environment for that case. In this thesis, preconditioning are the moisture and atmospheric instability conditions over an area prior to the initiation of heavy precipitation.

Model Validation

The mission of using a realistic model set-up is to take into account all possible affecting processes during past precipitating events and to enable the identification of possible model uncertainties in near-to-operational configurations. Verification and validation methods are used to assess the model accuracy. All of them give a quantification of how close the simulations are to observations. The observational data sets used for validation can be found in Section 5.3. The metrics and validation scores employed in this thesis are summarized in the following table.

Table 5.1: Thresholds for plausibility test used for the post-processing of the flux data.

Mean	Standard Deviation
$\mu = \frac{1}{N}\sum_{i=1}^{N} x_i$	$\sigma = \sqrt{\frac{1}{N}\sum_{i=1}^{N} \lvert x_i - \mu \rvert^2}$
Mean Bias	Mean Absolute Error
$BIAS = \frac{1}{N}\sum_{i=1}^{N}(y_i - x_i)$	$MAE = \frac{1}{N}\sum_{i=1}^{N} \lvert y_i - x_i \rvert$
Root Mean Square Error	Correlation Coefficient
$RMSE = \sqrt{\frac{1}{N}\sum_{i=1}^{N}(x_i - y_i)^2}$	$R = \frac{\frac{1}{N}\sum_{i=1}^{N}(y_i - \mu_y)(x_i - \mu_x)}{\sigma_y \sigma_x}$
Agreement Index	Cumulative Distribution Function
$AI = 1 - \frac{\sum_{i=1}^{N}(y_i - x_i)^2}{\sum_{i=1}^{N}\left(\lvert y_i - \mu_y \rvert + \lvert x_i - \mu_y \rvert\right)^2}$	$F_X(x) = \int_{-\infty}^{x} f_x(t)\, dt$

The Structure Location Amount (SAL) precipitation verification method (Wernli et al., 2008)

The SAL method is an object-based verification method for quantitative precipitation forecasts. With this representation, besides the amount and intensity of precipitation, the accuracy in the representation of the location can be assessed. The SAL method identifies the different precipitation structures by defining a threshold value depending on the maximum precipitation in the field $R* = fR^{max}$. Then, three different metrics quantify how similar the precipitation field is to that of a reference run, these are the amount (obtained as the sum precipitation of all found structures), the location (depending on the centre of masses of the different structures) and the structure (dependent of the ration of the volume of the found objects. Values of zero for the three quantities would imply a perfect similarity between the compared data sets. As indicated in the concept (Fig.5.1).

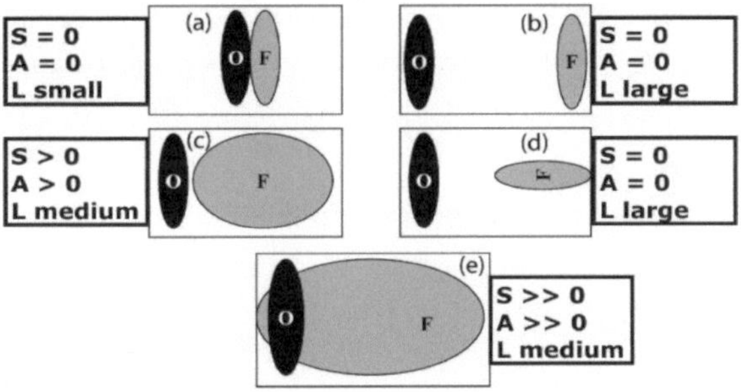

Figure 5.1: Schematic example of various comparisons between forecasts (F) and Observations (O). It is assumed that precipitation rates are uniform and the same in all objects. When the two objects have the same amount A=0, if the structure is the same S=0 and if the location is the same L=0. Obtained from (Wernli et al., 2008).

5.2 The 7 km, 2.8 km and 500 m Moisture Nudging Experiments

Two sets of simulations are performed

1. A seasonal simulation with COSMO in CLimate Mode (COSMO-CLM) in its version 5.00 covering the HyMeX SOP1 period from 1-Sep to 20-Nov

2. An event-scale simulation with COSMO in NWP mode (version 5.01) of the HyMeX-IOP6 from 22-Sep-2012 to 25-sep-2012

The main characteristics of the simulations are summarized in Tables 5.2 and 5.3. For both sets of simulations reference runs without any observations are produced (CTRL) to be compared to runs with assimilated data (AS). A nesting strategy is followed for dynamical downscaling of the simulations in which the simulations on the 7 km grid length type are forced by Integrated Forecast System (IFS) data, the 2.8 km runs are forced with the reference runs on the 7 km grid (CTRL-7) and the 500 m runs are forced by the CTRL-2.8 simulations.

Seasonal simulations - The HyMeX-SOP1

These simulations span the period 1-Sep to 20-Nov 2012. They are produced in a 7 km and a 2.8 km grid length using COSMO in Climate Mode (COSMO-CLM). The 7 km runs, cover large areas of Northern Europe, the Atlantic Ocean, Northern Africa and the whole Mediterranean basin. The 2.8 km simulations cover three climate regions the Iberian Peninsula (IP), France (FR) and Italy (IT) within the 7 km simulation domain as shown in Fig. 5.2. The main difference between using a 7 km grid with respect to using a 2.8 km are the increased number of vertical levels, the improved representation of the model surface and orography and the switching off of the deep convection parameterization scheme (see Table 5.2).

Table 5.2: Performed simulations with COSMO-CLM (seasonal simulations). CTRL stands for the reference runs without applying the Nudging Scheme. AS stands for the simulations containing nudged ZTD and the different horizontal resolutions are denoted by 7 and 2.8.

Name	Domain	Obs.	Levels	Resol.	Convection	Forcing
CTRL-7	Medit	-	40	7 km	Tiedtke Deep	IFS
AS-7	Medit	GPS	40	7 km	Tiedtke Deep	IFS
CTRL-SP-2.8	Iberia	-	60	2.8 km	Tiedtke Shallow	CTRL-7
AS-SP-2.8	Iberia	GPS	60	2.8 km	Tiedtke Shallow	CTRL-7
CTRL-IT-2.8	Italy	-	60	2.8 km	Tiedtke Shallow	CTRL-7
AS-IT-2.8	Italy	GPS	60	2.8 km	Tiedtke Shallow	CTRL-7
CTRL-FR-2.8	France	-	60	2.8 km	Tiedtke Shallow	CTRL-7
AS-FR-2.8	France	GPS	60	2.8 km	Tiedtke Shallow	CTRL-7

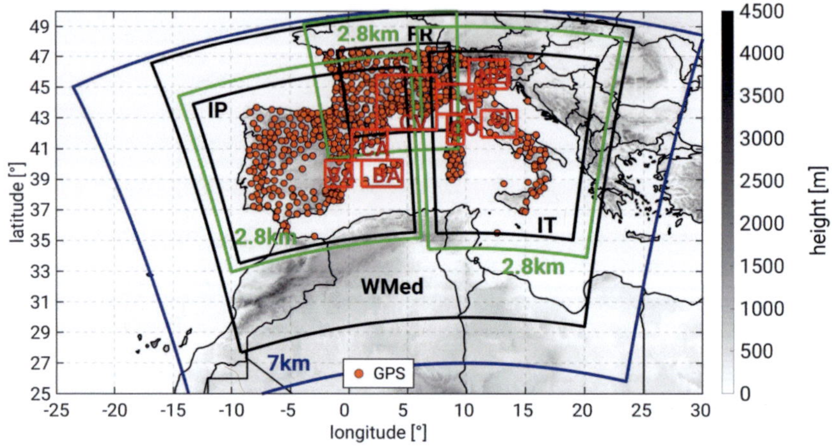

Figure 5.2: Simulation domains of the 1-Sep to 20-Nov simulations (SOP1 simulations). Blue boxes denote the extension of the 7 km rotated grid. Green boxes represent the three 2.8 km simulation domains. The HyMeX target areas are included in the red boxes. The scatter points show the location of the GPS receivers. The black boxes denote the four four climate investigation areas.

The coverage of the GPS observations used for nudging is shown in Fig. 5.2. New data enters the nudging cycle every 10 minutes. To the author's knowledge, it is the only experiment performed to date, nudging exclusively column moisture observations with such a high temporal resolution throughout a complete season (autumn), on the western Mediterranean, using merged reports from more than 20 European GPS networks.

Statistical analysis of precipitation, humidity and convective-related quantities are performed over the four climate regions, together with smaller, localized, target areas for heavy precipitation (the 8 HyMeX target areas). In this regard different temporal analysis is carried out, either for the whole period or for specific case studies (IOP6, IOP16).

- The IWV-precipitation relationship method

A novel method has been developed in the course of this thesis and used to characterize heavy precipitation during the HyMeX-SOP1. The purpose of the said method is to identify sub periods of heavy precipitation within the SOP1 and to relate them with IWV fluctuations taking place before precipitation onset over the affected areas. The *IWV-precipitation relationship* method aims at addressing relevant questions such as how often do IWV increases occur prior to precipitation initiation. Also, what is the order of magnitude of these IWV increments and how fast do they happen, ultimately correlating the strength of the said IWV growths with the intensity of each event. The method consists of the following steps:

1. Hourly IWV and precipitation are spatially averaged over each of the eight HyMeX target areas

2. HPEs in the spatially averaged temporal evolution of precipitation are identified as subperiods of hourly spatially-averaged precipitation larger than 0.15 $mm\ h^{-1}$ reaching, at least once, 0.4 $mm\ h^{-1}$. The timely and spatially averaged precipitation for each event is computed.

3. For each found HPE, a preconditioning period is identified as the time interval prior to precipitation onset where monotonic variations of IWV took place (increase or decrease) without any spatially averaged precipitation larger than 0.15 $mm\ h^{-1}$. In other words, the preconditioning period is the time between the onset of precipitation for each event and the hour when the IWV temporal evolution changed from a decrease to an increase or vice versa.

4. The variation of IWV is obtained as the change in IWV between the start of the preconditioning period and the start of precipitation.

5. The events are classified in types of degrees of synoptic forcing (as explained earlier in the chapter) based on the hourly τ values which were most frequent during the precipitation time window in the reference runs.

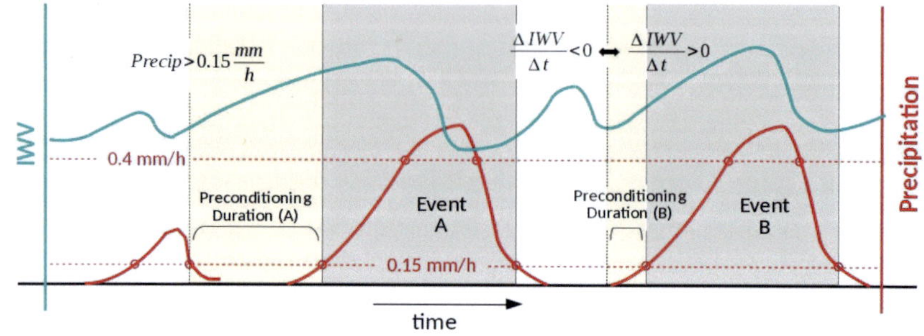

Figure 5.3: Concept of the IWV-precipitation causality method. Precipitation over the area is assumed to be sufficiently large for values of spatially averaged hourly precipitation larger than 0.15mm/h. If, in addition, precipitation surpasses 0.4mm/h, that period is considered as one event. Preconditioning is understood here as the period of monotonic IWV increases or decreases prior to precipitation onset. The duration of the preconditioning time interval depends on whether some precipitation over the area exists and on the change of sign of the monotonic IWV variations. Hence the preconditioning period is the time between the last transition from an IWV increase to a decrease (or vice versa) or since the last precipitation surpassing the 0.15 $mm\ h^{-1}$ threshold and the start of precipitation.

6. All identified HPEs among all areas are considered in statistical and physical analyses

After identifying the events of the period, these are put together for each simulation type for calculation of statistical quantities. At this point a heavy precipitation event will be defined as an event falling in the percentile-90 of the distribution.

Event-scale simulations - The HyMeX-IOP6

The HyMeX-IOP6 was dedicated to represent an HPE taking place during the night of 24-Sep-2012 over broad regions of southern France, north-eastern Italy over the Alps, and the Veneto and Friuli regions. It was selected for in-depth study of the relevance of the vertical distribution of humidity by combining Zenith Total Delay (ZTD) data and radiosondes.

Table 5.3: Performed simulations with COSMO for IOP6. Likewise the seasonal simulations of SOP1, in the naming of the different simulations CTRL stands for the reference run and the grid lengths are denoted by the number at the end 7, 2.8 and 500 m. The runs including the nudging of ZTD or operational sounding data are denoted by GPS and RAD, respectively.

Name	Obs.	Levels	Resol.	Convection	Turbulence	Forcing
CTRL-7	-	40	7 km	Tied. Deep	1D	IFS
AS-GPS-7	GPS	40	7 km	Tied. Deep	1D	IFS
AS-RAD-7	sondes	40	7 km	Tied. Deep	1D	IFS
AS-GPS-RAD-7	GPS & sondes	40	7 km	Tied. Deep	1D	IFS
CTRL-2.8	-	50	2.8 km	Tied. Shallow	1D	CTRL-7
AS-GPS-2.8	GPS	50	2.8 km	Tied. Shallow	1D	CTRL-7
AS-RAD-2.8	sondes	50	2.8 km	Tied. Shallow	1D	CTRL-7
AS-GPS-RAD-2.8	GPS & sondes	50	2.8 km	Tied. Shallow	1D	CTRL-7
CTRL-500	-	80	500 m	-	3D	CTRL-2.8
AS-GPS-500	GPS	80	500 m	-	3D	CTRL-2.8
AS-RAD-500	sondes	80	500 m	-	3D	CTRL-2.8
AS-GPS-RAD-500	GPS & sondes	80	500 m	-	3D	CTRL-2.8

The COSMO model in its NWP version is used in this case. The choice of the NWP version is desirable since for simulations of a few days (event-scale) adaptation of the climatic variables, such as ozone for instance is not needed and are considered constant (see Section 4.1). The atmospheric column moisture data (GPS-derived ZTD) are assimilated again, with a frequency of ten minutes. The same data set as in the seasonal simulations is used (see a description of the data set in Section 5.3). Regarding the operational soundings, three variables are nudged, temperature, dew point temperature and winds. These variables are the core variables assimilated in operational cycles.
The simulation of HyMeX-IOP6 spans the period 22-Sep-2012 00:00h to 25-Sep 2012 00:00h. They are produced in 7 km, 2.8 km and 500 m grids. The 7 km simulations are forced by IFS data and an analogous nesting strategy is followed for the 2.8 km and the 500 m grid simulations. In a like manner, when a 500 m resolution is reached, the number of levels is larger (80 levels), the refinement of the model surface is improved, and the convection parameteriza-

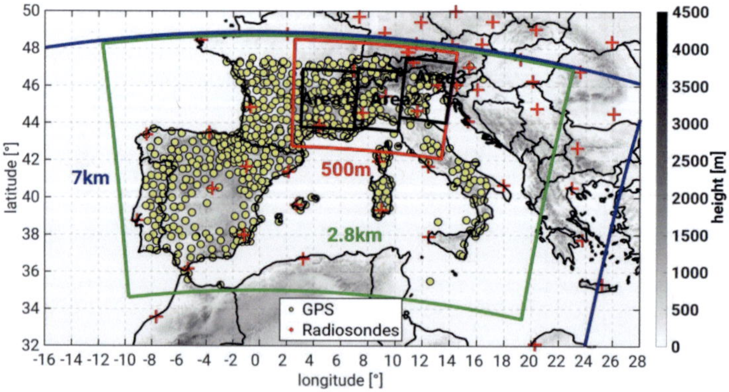

Figure 5.4: Simulation domains for the HyMeX-IOP6 event. Blue boxes denote the extension of the 7 km rotated grid. Green boxes represent the three 2.8 km simulation domains and red those of the 500 m grid. The three study regions, Area1, Area2 and Area3 are defined in black boxes The scatter points show the location of the GPS receivers and the crosses the location of the soundings.

tion scheme is completely switched off. Additionally, the subgrid-scale turbulent processes for this high-resolution simulations use a 3D closure based on the Turbulent Kinetic Energy (TKE). This is done since at these model resolutions, the boundary layer approximation (horizontal homogeneity of variables and fluxes) is no longer valid (it is applicable when the horizontal scales of motion are much larger than the vertical). In which case, a 3D turbulent diffusion of mass and energy should be applied. Table 5.3 summarizes the most relevant characteristics of the simulations.

5.3 Observational Data Sets

The 5 min frequency GPS-derived ZTD and IWV data sets

The outstanding, GPS-derived ZTD and IWV products made available to the HyMeX community, feature a very high temporal resolution. The ZTD product is made available up to a frequency of only 5 minutes and has a large coverage of the European WMed countries (see Figure 5.5). The data set has been produced at the *Laboratoire de Recherche en Geodesie* (LAREG) of the *Institut National de L'Information Géographique et Forestière* (IGN) (Bock

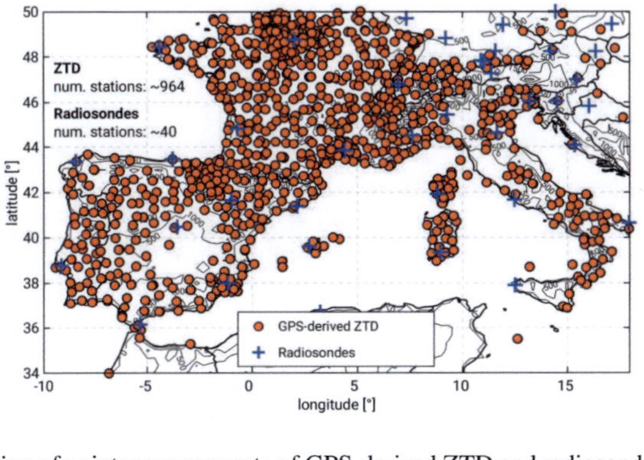

Figure 5.5: Location of point measurements of GPS-derived ZTD and radiosondes. GPS-ZTD has a maximum temporal resolution of 5 minutes.

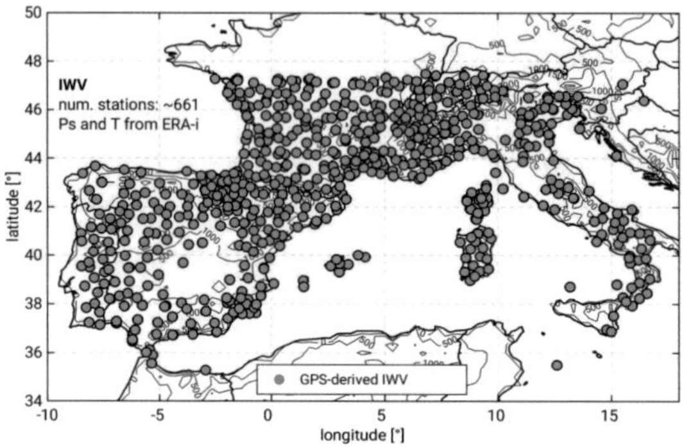

Figure 5.6: Location of point measurements of GPS-derived IWV (right). The GPS-IWV product is derived from the GPS-ZTD measurements using the station surface pressure and temperature from ERA-interim. The GPS-IWV is used in hourly or daily means.

et al., 2016), and what renders this data set unique is that it has been produced by commonly post-processing GPS measurements from 25 different networks in Europe. To this end, the GIPSY/OASIS II post-processing algorithm was applied in Precise Point Positioning mode to obtain the phase measurements. Tropospheric delays were obtained by time-varying Zenith Hydrostatic Delays (ZHDs), Zenith Wet Delays (ZWDs) and horizontal gradients along with their respective Vienna Mapping Function 1 (VMF1) mapping functions (Boehm et al., 2006). The ZWD parameters and horizontal gradients were modelled as random-walk processes with a 5 min time resolution and corrections to the a priori values were estimated during the data processing. The final ZTDs were obtained by combining the a priori ZHD and ZWD values and the estimated ZWD corrections with formal errors for the GPS ZTD of typically 1.5–5 mm (Bock et al., 2016). During the post-processing procedure quality assessment and data screening was applied to the ZTD calculations. Among others, restrictions to suspicious data depended on range checks and outliers checks for σ_{ZHD}, ZTD and σ_{ZTD}.

The GPS-ZTDs give a measure of the total delay endured by the emitted signal from the GPS satellites as it propagates through the troposphere. The ZTD measured by a single receiver can be split into the delay endured due to the interaction with the water vapour molecules (ZWD), exclusively, and the other elements of the troposphere. The latter is called dry zenith delay or Zenith Hydrostatic Delay (ZHD) and is derivable from the hydrostatic equation and the pressure at the GPS station location (p_s). Provided ZWD, yields information on the total delay corresponding to water vapour, it is proportional to the total column water vapour amount, or IWV. To derive IWV, the temperature the GPS-station location and height must be known. The following set of equations provides the relation among the four introduced quantities (Bevis et al., 1992).

$$ZTD = ZWD + ZHD \tag{5.1a}$$

$$ZHD = \frac{2.2765 \cdot p_s}{1 - \varepsilon} \tag{5.1b}$$

$$IWV = \frac{1}{R_v \cdot (k_s + k_l / T_m)} \cdot ZWD \tag{5.1c}$$

Where p_s is the pressure at the GPS station in [hPa], $\varepsilon = 2.66 \cdot 10^{-3} cos(2\phi) - 2.8 \cdot 10^{-7}$ is a factor depending on the station latitude (ϕ) and its height (z), R_v is the gas constant for water vapour and $k_s = 2.21 \cdot 10^{-4}$ and $k_l = 3.74$ are refractivity coefficients for the GPS signal propagation.

Thanks to this conversion, a second data set is made available to the HyMeX community, containing the reports of the IWV measured in the Zenith direction at the location of the GPS receiver. This data set is employed for validation in this thesis with a temporal evolution of 1h. The surface pressure and the temperature at the GPS-station location needed for the IWV conversion have been obtained from ERA-interim reanalyses data, interpolated bi-linearly to the station location (Bock et al., 2016). The spatial distribution of the point measurements is shown in Figs. 5.5 and 5.6.

In order to quantify the accuracy of the post-processed ZTDs, Bock et al. (2016) compared the LAREG GPS-ZTD product and the EUMETNET EIG GNSS water vapour programme (E-GVAP) ZTD product, which is assimilated operationally into NWP systems in Europe, assessing no significant biases. Regarding the converted IWV product, Bock et al. (2016) validated it against radiosonde measurements obtained during the HyMeX-SOP1 campaign, obtained from three different manufacturers, Modem M10, Vaisala RS92 and Graw DFM-09. After applying a height difference correction, the authors

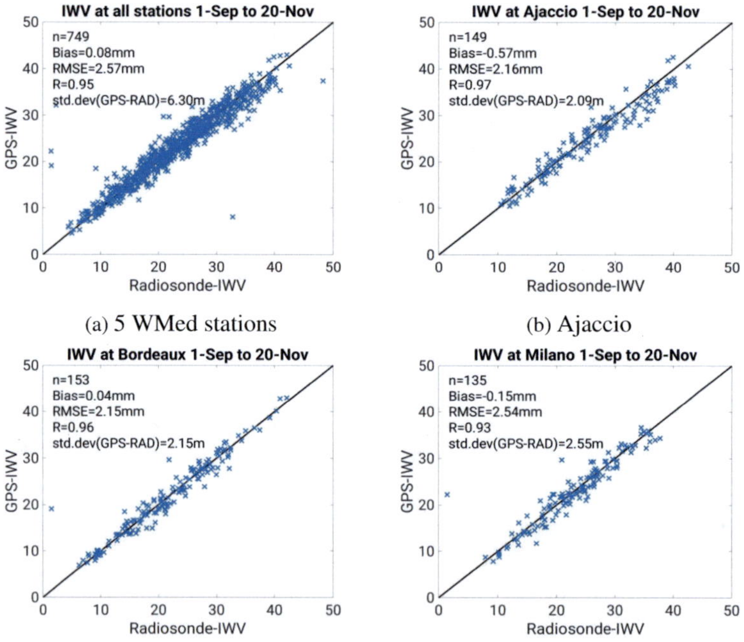

(a) 5 WMed stations (b) Ajaccio

(c) Bordeaux (d) Milano

Figure 5.7: Scatter plots for the IWV validation of GPS against radiosondes. Five stations within the WMed are selected and all available soundings within the 1-Sep to 20-Nov-2012 period are considered.

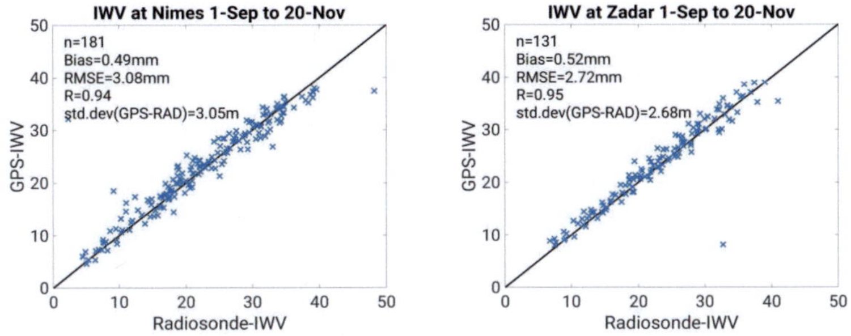

(a) Nimes (b) Zadar

Figure 5.8: As Fig. 5.7 for Nimes (France) and Zadar (Croatia).

found a significant overall bias of $\pm 3\%$ in IWV which is quite small compared to other experiments in the area (Bock et al., 2016).

The IWV data product of the HyMeX community has been validated separately against 5 radiosonde stations distributed between France and Croatia in the period 1-Sep to 20-Nov-2012 (see Figs. 5.7 and 5.8). The five selected stations are Ajaccio and Milano in Italy, Bordeaux and Nimes in France and Zadar in Croatia. The good agreement between data sets confirms the results by Bock et al. (2016), with a found correlation of 0.95 between the radiosondes and GPS, a mean bias of 0.11 mm and a RMSE of 2.61mm.

The Moderate Resolution Imaging Spectroradiometer (MODIS)

MODIS is utilized in order to study the IWV content of the atmosphere. It is obtained from radiance measurements performed by the imaging spectrometers on board the polar-orbiting platforms Terra and Aqua (crossing the Equator at 13:30h local time). IWV is derived by applying an algorithm to IR spectral radiances retrievals obtained with clear sky conditions during either day or night-time (Gao and Kaufman, 2003). The result is a highly spatially resolved gridded data set (5x5 km) of IWV, mapping the whole Earth's surface every one or two days. The employed data set for this thesis is a daily average of both Terra and Aqua observations made available on a grid spacing of 1°x1°.

The operational soundings

The operational soundings during the SOP1 field campaign, are also available for use within the HyMeX community. They are supplied in TEMP format by MétéoFrance and issued to the community by the teams ESPRI/IPL and SEDOO/Observatoire Midi-Pyrénées. They contain the following parameters at the most significant levels, dew point temperature, wind speed, geopotential height, air pressure, wind direction and speed and air temperature. In average they contain ca. 30 levels between the surface and the 300hPa level which in average implies one measurement every 250m. Most of the stations, launched the sounding balloons every 12h whilst some even have a temporal resolution of 6h, depending on the time period. In total, 40 stations fall within the WMed investigation area and the spatial distribution can be seen in Figure 5.5.

Rain Gauges

The Rain Gauges data set is provided as well by MétéoFrance and the HyMeX team. In the latest available version, accumulations for 1h, 6h and 24h are available, where the date of the file refers to past accumulation periods. This data set results from merging four European networks, Spain, Catalonia, France and Italy. Consistency checks among the different accumulation times are applied as well as quality control checks. The data set enjoys a dense coverage of the European countries with a total number of gauges close to 4800, within the WMed domain (see figure 3.5).

The NOAA CPC Morphing Technique (CMORPH)

This product provides precipitation estimates during the investigation period in a gridded data set derived from microwave observation instruments placed in different low orbiter satellites. Microwave measurements from the Special Sensor Microwave Imager (SSM/I) aboard the Defence Meteorological Satellite Program satellites (DMSP-13, -14 and -15), the Advanced Microwave Sounding Unit-B (AMSU-B), on-board the National Oceanic and Atmospheric Administration spacecraft's (NOAA-15, -16, -17 and -18 spacecraft's), the Advanced Microwave Scanning Radiometer-Earth Observing System (AMSR-E) of the Aqua spacecraft and the passive Microwave Imager of the Tropical Rainfall Measuring Mission satellite (TMI-TRMM) are combined by means of the Morphing Technique (Joyce et al., 2004). This procedure makes use of infrared derived motion vectors to propagate the microwave imager's features to locations where no microwave data were obtained at a specific time (Stampoulis et al. 2013). This renders the technique flexible regarding its applicability to any microwave satellite source. In addition, it offers a wide coverage (60°S to P 60°N), and a high temporal and spatial resolution, 30 minutes and 8 km at the Equator. The data set is produced by the Climate Prediction Center (CPC) of the National Weather service (NWS) of the USA and covers the period between 1998 and 2015.

The Global Land Evaporation Amsterdam Model (GLEAM) product

GLEAM is a daily gridded terrestrial evaporation (evapotranspiration) data set using, mainly, remote sensing data (Martens et al., 2017). Four modules simulate the soil processes relevant to calculate the contributions to land evapotranspiration defining the algorithms of the GLEAM product. Total evapotranspiration is computed as the sum of Transpiration (E_t), bare-soil evaporation (E_b), Snow sublimation (E_s), open-water evaporation (E_w), interception loss (E_i) and snow sublimation (E_s).

$$E = E_t + E_b + E_w + E_i + E_s \tag{5.2}$$

Input variables for these four modules are obtained where possible from gridded satellite data. In the case of insufficient data, reanalyses (mostly from ERA interim) are used. Radiation data are obtained from the Clouds and the Earth's Radiant Energy System (CERES) (Wielicki et al., 1996). Precipitation is obtained from the TRMM Multisatellite Precipitation Analysis (TMPA) (Huffman et al., 2007). Air temperature is obtained from the Atmospheric Infrared Sounder (AIRS) (Aumann et al., 2003) aboard the Terra and Aqua satellites. The soil moisture is obtained from the L3 product of the Soil Moisture Ocean Salinity (SMOS) project (Jacquette et al., 2010). The GLEAM product used for validation of evapotranspiration over northern Africa (see Chap. 7.4) is a daily accumulated evapotranspiration product in a 0.25°x0.25° grid.

6 Column Moisture Nudging to Improve the Modelling of Atmospheric Moisture

Numerical weather prediction of deep moist convection has shown deficiencies linked to errors in the simulation of the atmospheric moisture distribution (Steinke et al., 2014; Chazette et al., 2015b; Khodayar et al., 2016b). Added value in heavy precipitation forecasts can be gained by improving the humidity distribution in weather and climate models, especially through data assimilation of very frequent humidity measurements.

In this chapter, the seasonal simulations performed with COSMO in CLimate Mode (COSMO-CLM) of the 1-Sep to 20-Nov period, as described in Chapter 5, are used to assess the accuracy of the atmospheric humidity representation. Two different model horizontal resolutions are employed. This study is carried out over three broad climate investigation areas of semi-arid (Iberian Peninsula, IP) and wet Mediterranean climates (France, FR and Italy, IT) as well as over the eight local HyMeX target areas (see section 3.2) to allow for an assessment of the atmospheric moisture distribution. The following scientific questions are set out:

Q.1 *How well is the atmospheric moisture distribution represented in COSMO-CLM for the HyMeX-SOP1, on a seasonal scale? Are there any particular periods or regions of large humidity model biases?*

Q.2 *What is the impact of nudging column atmospheric moisture on the representation of humidity? How does the continuous GPS-ZTD nudging affect its spatial distribution, the timing, the represented humidity extremes, the moisture transport and the diurnal cycle?*

Q.3 *What are the differences and similarities of the water vapour representation between the 7 km and the 2.8 km grid settings?*

6.1 Spatio-temporal Representation of Integrated Water Vapour

The IWV fields are analysed first given the proportional relation between Zenith Total Delay (ZTD), the nudged variable, and IWV. Besides, characterizing the spatio-temporal distribution of IWV during the study period will provide useful information of large water vapour transport at the large ($> 2 \cdot 10^3$ *km*) and convective scales ($> 10^0$ *km*).

HyMeX-SOP1 spatial distribution of IWV

During the Special Observation Period 1 (SOP1), extreme IWV was represented along the eastern Spanish coast, the Balearic Islands, southern France along the Rhone valley, the islands of Corsica and Sardinia and the north eastern Italian regions around the Po valley (see Fig. 3.4). The seasonally aver-

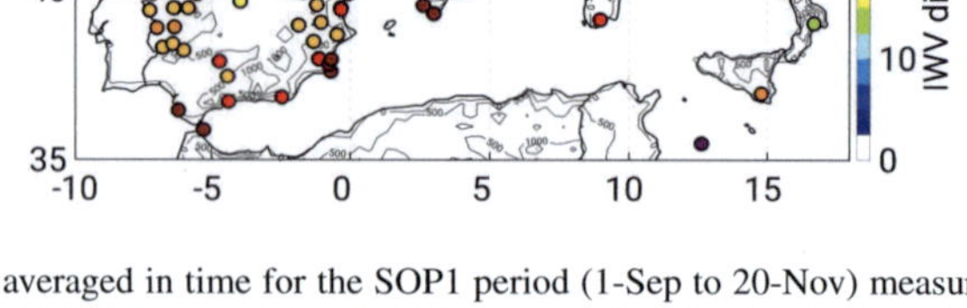

Figure 6.1: IWV averaged in time for the SOP1 period (1-Sep to 20-Nov) measured by the GPS-derived IWV data set of the HyMeX network (see Section 3.2 for a full description of the data set).

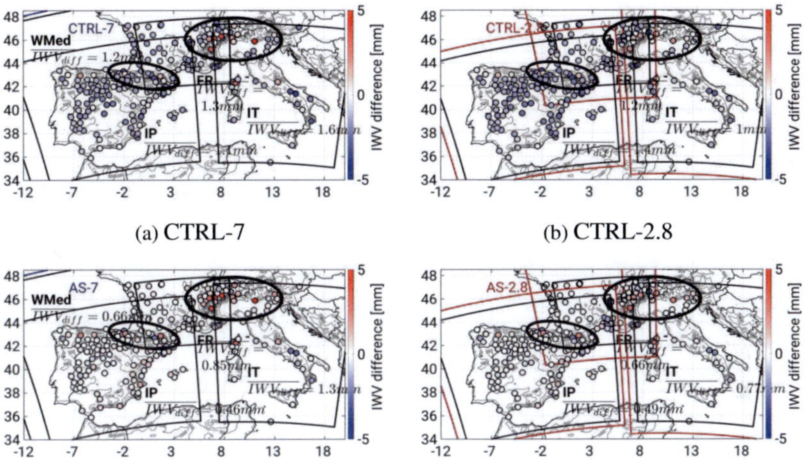

(a) CTRL-7

(b) CTRL-2.8

(c) AS-7

(d) AS-2.8

Figure 6.2: Daily IWV differences (GPS-COSMO) averaged for the period 1-Sep to 20-Nov, at the location of the GPS stations. An interpolation of the closest model grid points is performed to the location of the GPS station. GPS measurements with missing daily IWV data are dismissed.

aged IWV values measured by the Global Positioning System (GPS) over the aforementioned regions ranged between 25 mm and 30 mm. These regions are prone to heavy precipitation during this time of the year (Lionello, 2012; Ducrocq et al., 2014). On the contrary, lower IWV amounts are present over elevated terrain such as the Alps, the Pyrenees, and over the northern Iberian plateau (of about 12 mm in the seasonal mean).

In order to assess the accuracy of the representation of seasonally averaged IWV by COSMO-CLM, the differences against the GPS-derived IWV are shown in a spatial distribution at the location of the GPS stations in Figure 6.2, These differences are obtained by interpolating IWV, as simulated by COSMO-CLM, from the nearest grid points to the station location. The Mean Absolute Error (MAE) quantifies the departures between the model and observations (a description of the MAE is included in Chapter 5). Relevant errors in the spa-

Table 6.1: Spatially averaged bias of the stations within the three climate regions (IP, FR and IT) of 24hly IWV differences, as shown in figure 6.2.

Mean Abs. Error (mm)	CTRL-7	AS-7	CTRL-2.8	AS-2.8
IP	1.1	0.5	1.4	0.5
FR	1.3	0.9	1.2	0.7
IT	1.6	1.3	1	0.8

tial distribution of IWV seasonal means (1-Sep to 20-Nov) exist between the GPS observations and the CTRL runs of both resolutions (7 km and 2.8 km). Overall, the model showed a wet bias in both the 7 km and the 2.8 km configurations, albeit particular point locations over mountain regions (the Alps and Corsica) showed a too dry atmosphere (see Fig. 6.2). The mean absolute error was reduced over all climate regions for the AS-7 and AS-2.8 simulations, with the exception of IT in the AS-7 runs. This is due to the low number of GPS stations with enough data to perform the season average along the Italian peninsula and the large IWV underestimation over elevated stations of the Alps. The mean absolute error over the climate regions ranges from 1 *mm* to 1.6 *mm* in the CTRL runs to 0.5 *mm* to 1.3 *mm* in the AS runs, depending on the resolution and selected area (see Tab. 6.1). The best improvement brought by the column moisture nudging took place over the Iberian Peninsula (IP) for both grid length types, and the worst, as stated before, over Italy. In addition to the errors stemming from the Alpine stations and the lower GPS coverage over the peninsula, Italy is less impacted by the nudging of the column atmospheric moisture given it is located downstream of southwesterly flows, typical for this time of the year. These flows traverse regions of no GPS coverage such as northern Africa and the Mediterranean Sea.

In spite of the generalized wet bias existing in CTRL-7 and CTRL-2.8 and its correction in AS-7 and AS-2.8, particular stations over the mountain regions of the Alps, the Pyrenees and Corsica show persistent point values of a too dry atmosphere. They are related to the misrepresentation of the surface height in COSMO-CLM, crucial for IWV representation. Given nearly 50% of the total atmospheric moisture is contained within the first 1.5 km (Munn et al., 2003) even small discrepancies in the height of the integrated water vapour column bring large differences in the modelled IWV (Bock et al., 2007). These positive deviations can be as large as 5 mm in the CTRL-7 and AS-7 runs and somewhat lower (3 mm) in the CTRL-2.8 and AS-2.8 runs. The nudging was not able to correct them for two reasons. First, GPS-ZTD observations only enter the nudging cycle if their distance difference to the real station height is shorter than 100 m for extrapolation (i.e. if the model surface height is lower than 100 m with respect to the GPS station height) or 400 for interpolation (in the case of the COSMO-CLM orography being higher than the station height). In total, 32 stations are dismissed from the nudging procedure for this reason in AS-7 and 31 in AS-2.8. Second, under such large height differences, any potential improvement caused by the nudging, for instance due to the influence of remote ZTD observations is negligible in comparison to the errors introduced by the height discrepancies. Finally, it is worth highlighting that the positive biases in AS-2.8, compared to AS-7, are smaller due to the shorter height differences between the real station and the modelled heights. This demonstrates the added value of using a high-resolution grid in the representation of IWV.

The humidity reduction takes place over the whole northern domain of the Western Mediterranean (WMed) for both types of grid lengths (7 km and 2.8 km). Figure 6.3 shows the relative variations of seasonally averaged IWV over the WMed between AS-7 and CTRL-7 for the 7 km grid. An analogue spatial distribution is found for the relative variations in the 2.8 km grid. The relative reductions are in the order of -3% to -8% for IWV over the WMed, with the largest mean variation taking place over FR and the weakest over IT for both resolutions (see table in Fig. 6.3). This can be explained by the larger number of stations deployed over FR as compared to IT and by the location of the Italian peninsula, downstream of Africa and the Mediterranean Sea, where few stations are deployed. At a local scale, when analysing the relative variations of

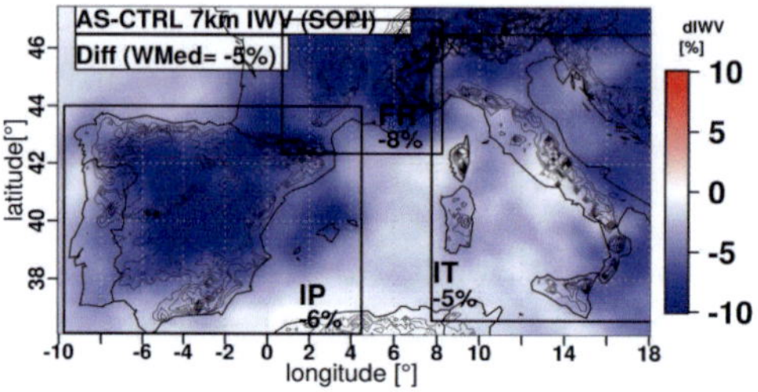

$\frac{AS-CTRL}{CTRL}$ [%]	7 km	2.8 km
IP	-6	-6
FR	-8	-7
IT	-5	-3
VA	-8	-8
BA	-6	-6
CA	-7	-7
CV	-8	-7
CO	-2	-2
CI	-3	-2
NEI	-7	-4
LT	-6	-4

Figure 6.3: Spatial distribution of timely averaged IWV relative differences [(AS-CTRL)/CTRL] in [%] for the 7 km simulations (left). Blue colours account for IWV in excess in the CTRL runs as compared to AS. The percentages of the seasonal relative differences, spatially averaged, are presented in the table (right) for the climate regions and the HyMeX target areas for the 7 km and the 2.8 km grids.

IWV over the HyMeX target areas, the order of the reductions lays within -2% and -8%. The largest reductions took place over Valencia (VA) and Cévennes-Vivarais (CV, -8%) and the weakest over Corsica (CO) and Central Italy (CI, -2%), in agreement with the relative variations of the parent broader climate regions.

In order to quantify better the changes in the spatial distribution of the timely averaged IWV, the Probability Distribution Functions (PDFs) over the climate regions for this variable are shown in Figure 6.4. The PDFs show the probability of a grid point showing a certain IWV value after averaging this variable in time for the whole period 1-Sep to 20-Nov. IP shows a broader range of most probable values between 15 mm to 30 mm. For FR, the most probable IWV is in the range 15 mm to 25 mm and for IT is within 22 mm to 28 mm. The larger IWV values in Italy arise from the consideration of more sea grid

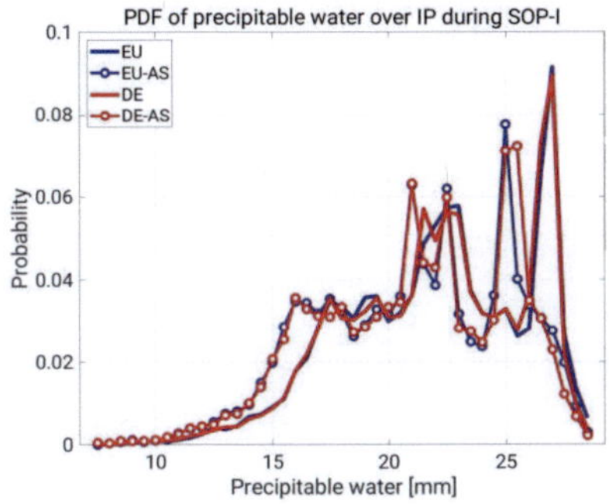

(a) IP

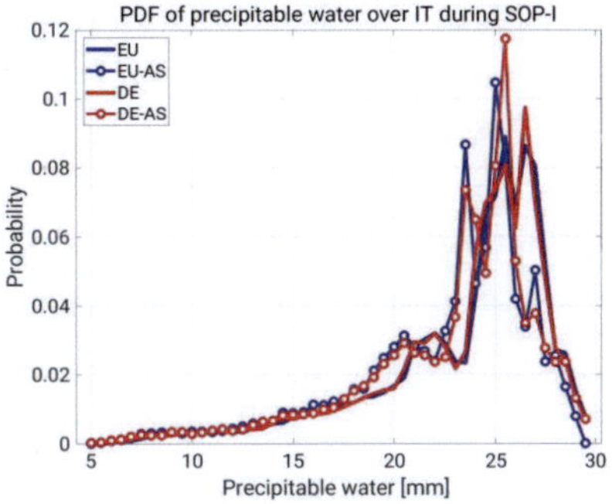

(b) IT

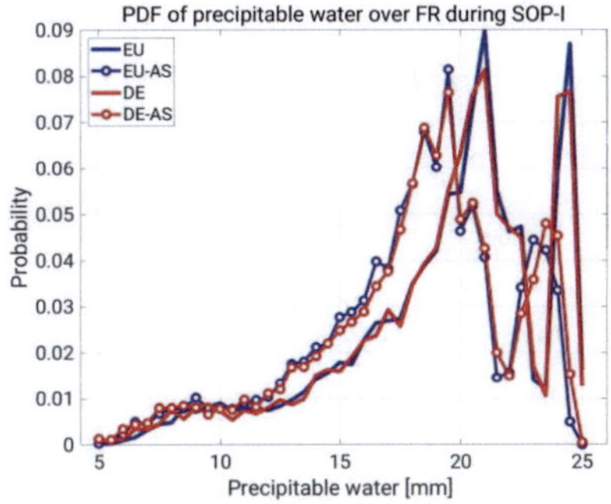

(c) FR

Figure 6.4: Probability Distribution Function (PDF) of the timely averaged (1-Sep to 20-Nov) IWV spatial distribution shown in Fig.6.3 for all simulations and the climate regions (IP, IT and FR). The ordinate axis, stands for the probability of a number of points showing a particular IWV value. All simulations have been upscaled to a coarser grid of 8km.

points in evaluating the PDFs. Remarkable differences in the representation of IWV PDF exist between the three climate regions. All areas after the nudging, show a shifting of about 2mm, toward lower values, of the most probable IWV amounts. This is in line with the humidity reduction shown by Figures 6.2 and 6.3. Over IP and FR a second maximum exists, at about 23 mm and 22mm, respectively, which also undergoes this shifting.

HyMeX-SOP1 temporal evolution of IWV

Regarding the temporal evolution of IWV, relevant biases are present in the CTRL-7 and CTRL-2.8 runs over the three climate regions, which are well corrected by the nudging of column atmospheric moisture. To illustrate this, Figure 6.5 shows the daily averaged IWV, temporal evolution over the climate region IT. The GPS-derived IWV is compared to COSMO-CLM IWV, interpolated to the location of the stations. The differences between the simulated and the observed IWV can be as high as 6 mm (for example at 11-Sep-2012, for both resolutions) in the CTRL runs. This holds for FR and IP. Over the smaller HyMeX target areas the discrepancies between the model and the observations are larger, up to 10 mm (11-Sep-2012 over Corsica, not shown). The timing of the IWV increases coincides the befallen of heavy precipitation in the investigation domain which is well represented by COSMO-CLM.

In order to quantify the degree of agreement between the observations and the different COSMO-CLM runs, the mean bias, the Root Mean Square Error (RMSE) and the Agreement Index (AI), of daily IWV in the period, are shown in Table 6.2 (see Chap. 5 for a description of these indices). All three metrics,

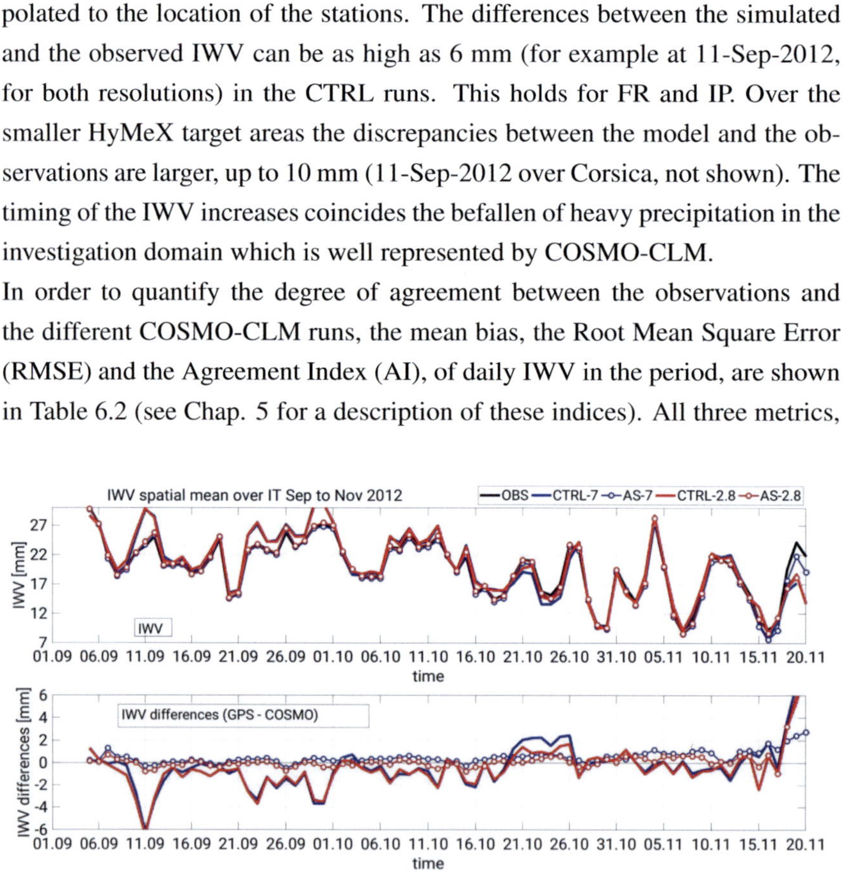

Figure 6.5: Temporal evolution of the spatially averaged daily IWV for the 7 km (blue) and 2.8 km (red) resolutions for the CTRL and AS runs.

CTRL-7	Bias	RMSE	AI
IP	-1.13	1.73	0.972
FR	-1.13	1.95	0.971
IT	-0.39	1.82	0.968
VA	-1.3	2.2	0.968
BA	-1.9	2.64	0.952
CA	-1.04	2.22	0.976
CV	-1.05	2.04	0.971
CO	0.17	3.69	0.916
CI	0.2	3.09	0.926
NEI	-0.62	1.88	0.972
LT	-0.55	2.46	0.958

CTRL-2.8	Bias	RMSE	AI
IP	-1.31	1.86	0.969
FR	-1.27	2.08	0.967
IT	-0.51	1.92	0.964
VA	-1.63	2.57	0.958
BA	-1.61	2.46	0.959
CA	-1.32	2.36	0.973
CV	-1.21	2.2	0.967
CO	-0.17	3.39	0.927
CI	-0.29	3.08	0.927
NEI	-1	2.1	0.966
LT	-0.49	2.46	0.957

AS-7	Bias	RMSE	AI
IP	0.22	0.39	0.999
FR	0.4	0.53	0.998
IT	0.49	0.74	0.994
VA	0.41	0.57	0.998
BA	0.2	0.63	0.997
CA	0.4	0.59	0.998
CV	0.44	0.59	0.998
CO	0.49	0.86	0.995
CI	0.66	0.79	0.995
NEI	0.56	0.66	0.996
LT	0.5	0.7	0.996

AS-2.8	Bias	RMSE	AI
IP	-0.04	0.33	0.999
FR	0.09	0.36	0.999
IT	0.13	0.87	0.992
VA	0.06	0.7	0.997
BA	0.04	0.56	0.998
CA	0.09	0.49	0.999
CV	0.09	0.41	0.999
CO	0.37	1.01	0.993
CI	-0.02	0.55	0.998
NEI	-0.28	0.53	0.998
LT	0.18	0.57	0.998

Table 6.2: Bias, RMSE and Agreement Index (AI) between COSMO-CLM and GPS-IWV for 24hly and spatially averaged IWV over the climate regions and HyMeX target areas. For a description of the metrics refer to Chap. 5.

show readily the good improvement brought by the moisture nudging with better scores for AS-7 and AS-2.8 over the climate regions and the HyMeX target areas. The negative bias in CTRL-7 and CTRL-2.8 ranging between -0.29 *mm* and -1.61 *mm*, depending on the area considered, are new evidence of the wet bias of the model. After the humidity nudging, the bias for AS-7 and AS-2.8 is within the range 0.2 *mm* to -0.7 *mm*. In this regard, the nudging of very frequent ZTD data is most beneficial in the 2.8 km grid, given the lower bias values for AS-2.8 as compared to AS-7. The best RMSE scores take place over IP, for both resolutions (RMSE of 1.73 in CTRL-7 and of 1.86 in CTRL-2.8) and the worst scores take place over FR (RMSE of 1.95 in CTRL-7 and of 2.08 in CTRL-2.8). After nudging column atmospheric moisture, the largest errors over FR are well corrected and both the AS-7 and AS-2.8 runs show a weaker impact over IT. (RMSE of 0.74 in AS-7 and of 0.87 in AS-2.8). Accordingly, for the HyMeX areas, the Italian regions CI and Corsica (CO) show the worst values for RMSE and AI, also after applying the nudging since they are particularly exposed to maritime flows, less affected by the nudging.
Contributing to the overall bias, there are four long-lasting periods of large discrepancies. The first biased period, between 6-Sep and 12-Sep, is apparent over all climate regions. It stems from very intense evapotranspiration over the northern Iberian Peninsula (8-Sep) and northern Africa (10-Sep) swept away by a slow moving westerly large-scale front. To validate the representation of this large moisture transport by COSMO-CLM, daily IWV measured by the Moderate Resolution Imaging Spectroradiometer (MODIS) is shown in Figure 6.6a (for a description of the used MODIS product, refer to Section 5). Figures 6.6c and 6.6e, represent the differences in daily IWV between MODIS and the CTRL-7 and AS-7 simulations, respectively (all model values have been upscaled by a linear interpolation method to the coarser resolution of MODIS). The 10-Sep is shown since it is the day in the period when the large moisture amount reaches FR with largest number of available pixels. Only the results on the 7 km are shown given the broader extent of the simulation domains, covering the Atlantic Ocean and Northern Africa. The IWV bias is likewise

present in the 2.8 km runs since they were forced by the CTRL-7 output. Over western France the spatially averaged differences are as large as -4.2 mm and

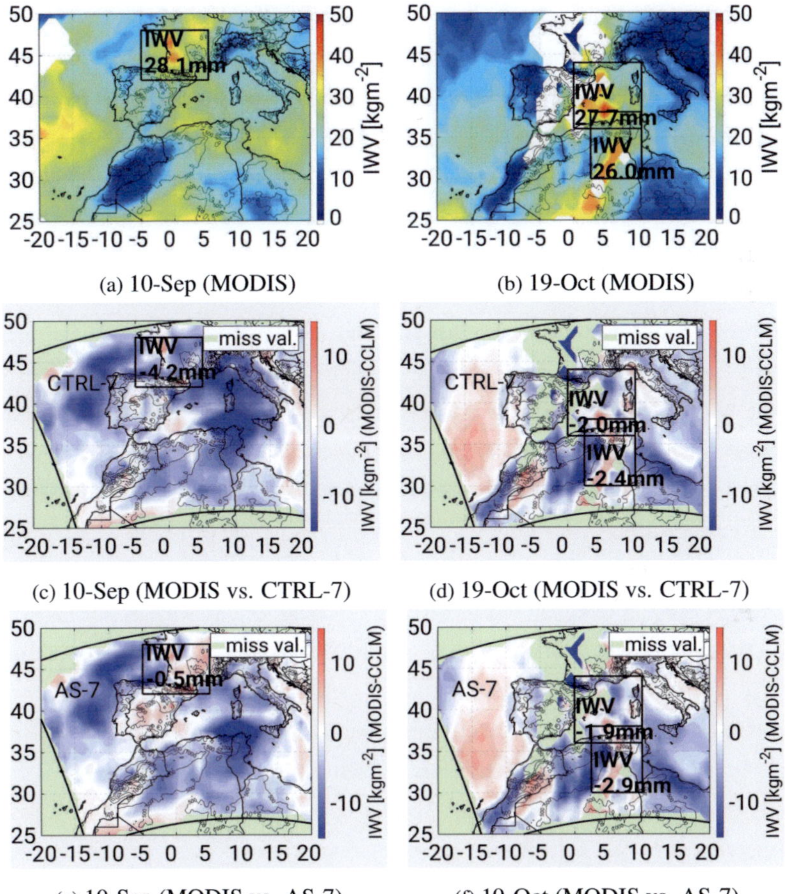

(a) 10-Sep (MODIS) (b) 19-Oct (MODIS)

(c) 10-Sep (MODIS vs. CTRL-7) (d) 19-Oct (MODIS vs. CTRL-7)

(e) 10-Sep (MODIS vs. AS-7) (f) 19-Oct (MODIS vs. AS-7)

Figure 6.6: Spatial distribution of daily averaged IWV as measured by the MODIS instrument aboard the Terra satellite (top) and difference plots of the same quantity against CTRL-7 and AS-7 (bottom). The dates, 10-Sep-2012 and 19-Oct-2012 are representative of the large IWV transport originating at the northern Iberian Peninsula and over north Africa, respectively. Both COSMO-CLM simulations have been upscaled to the MODIS grid of 1°x1°.

after nudging column atmospheric moisture, the wet bias is reduced to -0.5 mm and -8.9mm, respectively. Hence is the moisture nudging beneficial for improving the humidity amount over FR in the period.

The second period of large moisture biases, affecting IT only, takes place between 21-Sep and 01-Oct. The first two days of this period of large IWV wet biases stem from humidity evaporating over Corsica and Sardinia at midday on the 22-Sep-2012, and during the 25-Sep and on the 26-Sep over the Algerian coast and northern Tunisia. These moist air masses reach Central and south Italy one day after forced by a large-wave trough descending toward southern Spain, inducing very fast winds (see Annex for the comparison against MODIS daily IWV).

The third period occurs between 18-Oct and 22-Oct (over FR only) and arises from a quasi-stationary trough-ridge synoptic configuration inducing strong southerly winds and moisture advection from North Africa and the Spanish Mediterranean Sea on the 19-Oct (Scheffknecht et al., 2016; Lee et al., 2017). Figure 6.6b shows the spatial distribution of daily averaged IWV from MODIS on the 19-Oct-2012 and Figures 6.6d and 6.6f show the differences against the daily IWV represented by CTRL-7 and AS-7. A large IWV amount (of about 45 mm) is present over the western Mediterranean sea as well as across northern Africa, the latter is of ca. 40 mm. Both COSMO-CLM runs underestimate the IWV amount at these localized areas (with values ca. 10 mm lower) but overestimate IWV in their surrounding, hence explaining the negative differences in the Algerian and Mediterranean Sea squares.

Finally, the fourth period, between 12-Nov and 20-Nov (all climate regions) is characterized by the location of a small-wave cyclone over Tunisia followed by the arrival of a large-scale cold front pushing humidity from the Atlantic during the first two days (see Annex for the comparison of COSMO-CLM against MODIS).

The exposed humidity wet bias coincides past studies with the COSMO model (Schraff et al., 2008; Cress et al., 2012; Devidasrao, 2012).

IWV variability and diurnal cycle

Figure 6.7 shows the diurnal cycle of IWV, averaged for every hour in the day, over all climate regions for both the 7 km and the 2.8 km simulations. These results, demonstrate once more, the overall reduction of humidity throughout the WMed. The daily cycle scales with diurnal radiation, showing larger IWV due to enhanced evapotranspiration in the afternoon and larger water capacity of the air masses with warming, as expected from the Clausius-Clapeyron equation (Drobinski et al., 2016). The nudging works as to correct what seems to be a too weak diurnal cycle, decreasing the amount of IWV by 1.5 mm in the early morning and by only 0.5 mm in the midday and afternoon for both resolutions and the three climate regions.

In order to assess how the column moisture nudging affected the representation of IWV extreme values, the PDF of hourly IWV over land points is obtained. Figure 6.8 illustrates the impact for two climate regions (FR and IT, upper

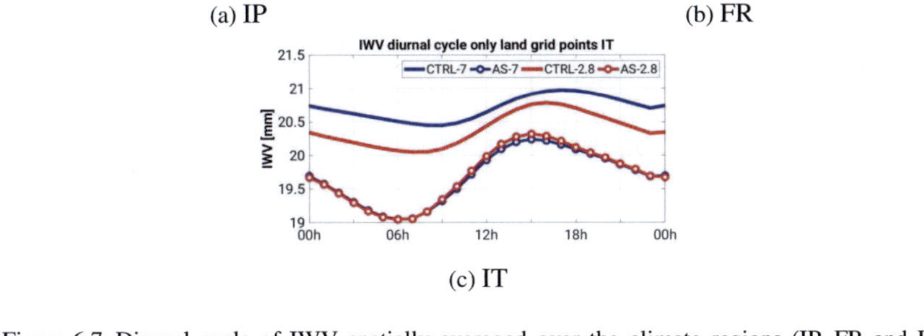

(a) IP (b) FR

(c) IT

Figure 6.7: Diurnal cycle of IWV spatially averaged over the climate regions (IP, FR and IT), timely averaged for the 1-Sep to 20-Nov period. Only land points are considered in this representation.

panel) and two HyMeX areas (CV and CO, lower panel), belonging to the aforementioned broader domains. Overall, the moisture nudging induced a reduction of the probability of maximum IWV. When considering the large climate domains, the probability reduction of extreme IWV was most clear for

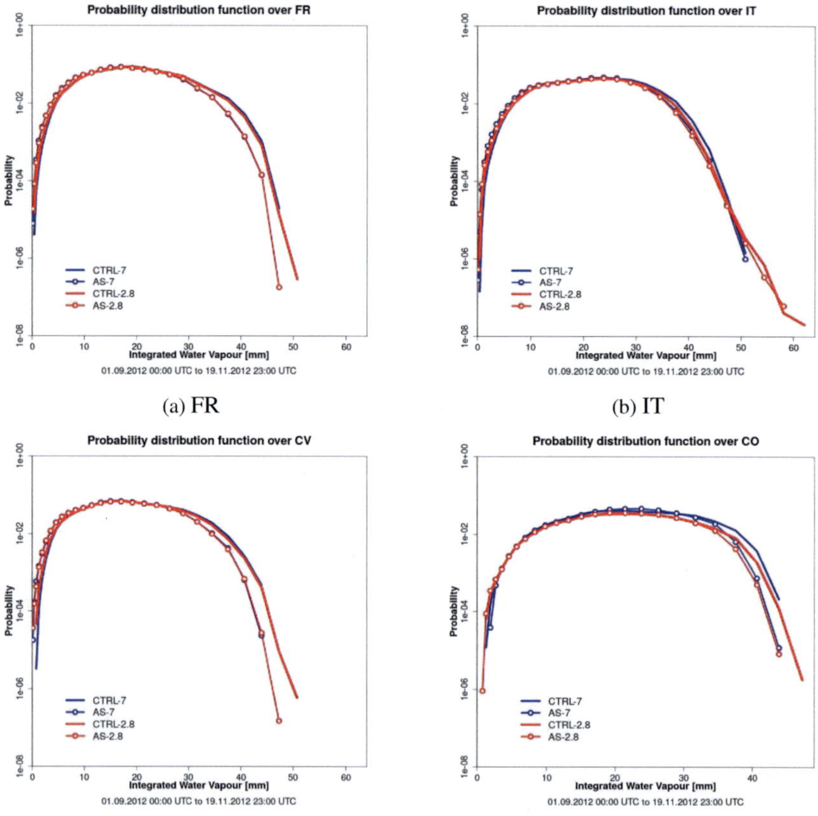

(a) FR (b) IT

(c) Cévennes-Vivarais (CV) (d) Corsica (CO)

Figure 6.8: Probability Distribution Function (PDF) of hourly IWV values represented by COSMO-CLM for the SOP1 period (1-Sep to 20-Nov) over two climate regions (FR and IT), and two HyMeX target areas contained within, these are, Cévennes-Vivarais (CV) in France the island of Corsica in Italy. Only land points are considered in this representation.

FR (AS-7 and AS-2.8 show less probability of large IWV values than CTRL-7 and CTRL-2.8), and smaller differences are seen for IT and IP. A larger impact of the nudging exists over FR given its larger coverage of GPS stations and its more humid climate by autumn. When considering the local target areas, the shift of probability for maximum IWV is clear over all of them. The Spanish stations (BA and VA) show clear differences between the CTRL and AS runs of both horizontal resolutions. CA, shows large discrepancies between the CTRL and AS runs of both resolutions for the whole range of IWV values (> 10 mm). Finally, the Italian (CO, CI, LT and NEI) and the French (CV) stations, experience a decrease in probability of extreme IWV for values larger than 35 mm.

6.2 The Effect on the Vertical Distribution of Atmospheric Moisture

The spatio-temporal representation of IWV improves after the nudging of column atmospheric moisture observations, especially in a model horizontal resolution of 2.8 km (better AI and RMSE scores in the temporal resolution and improvement of the spatial distribution over mountain regions). This section assesses how accurate the redistribution of the model humidity in the vertical direction is. This aspect is crucial to understand how convection will evolve, since the stratification of humidity is a determinant factor for the convective variables, possible inversions and decrease of mixing. Furthermore, as introduced in Chapter 4.2, the redistribution of the ZTD nudged information (integrated quantity), to the vertical profile is performed internally by COSMO-CLM based on the existing model profile at the time. Hence, it is expected that errors may arise from the artificial redistribution of the integrated information in the vertical direction.

In order to assess the accuracy of the vertical distribution of humidity, before and after nudging column atmospheric moisture, soundings from three operative stations providing data every 6h to 12h are compared to the differ-

ent COSMO-CLM runs. The three sounding stations, belonging to each of the broad climate regions, Barcelona in Spain, Nimes in France and Pratica di Mare in Italy, are selected for comparison. These stations cannot be considered representative of the whole climate domain, given the large variability of climate types within each of them. However using them gives an estimation of the improvement of the moisture nudging procedure for the vertical distribution of humidity at the station location. To date, radiosondes remains the backbone of atmospheric observation systems for profile sampling given the accuracy of this measurement technique, hence their election for model validation.
Figure 6.9, shows the RMSE between the soundings and COSMO for the three selected stations at all available hours. Between 50 and 80 soundings of day and night measurements were used for each station in the validation, depending on the considered station. The RMSE at 11 pressure levels is shown, where the sounding data has been interpolated vertically to those 11 levels and the COSMO-CLM specific humidity values are taken from the closest grid point to the station location. The comparison over the three stations shows an improvement of the humidity RMSE between 500 hPa and 950 hPa, for the AS-7

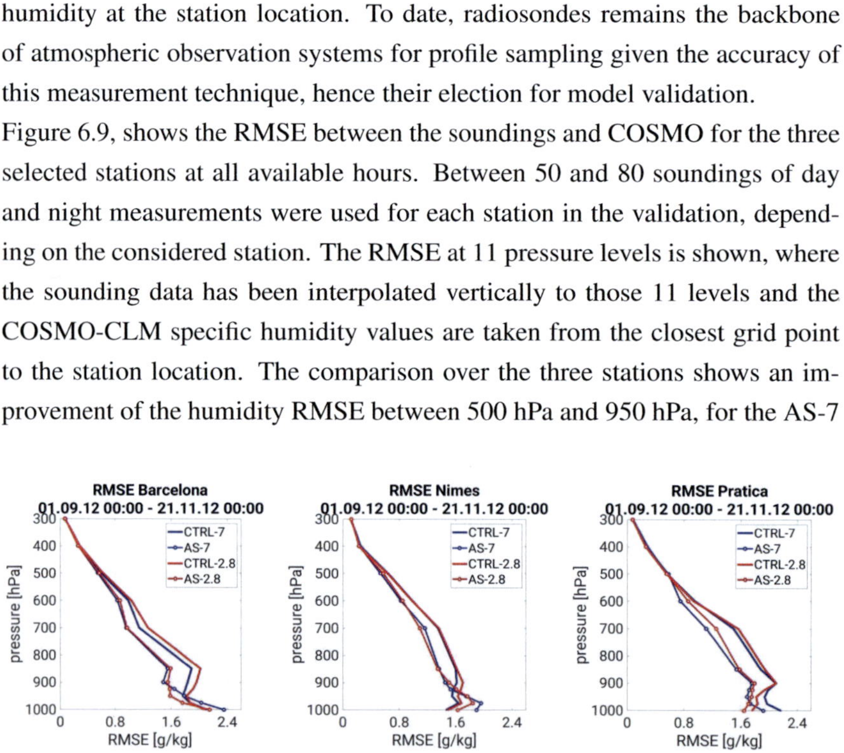

(a) Barcelona (Spain) (b) Nimes (France) (c) Pratica (Italy)

Figure 6.9: Vertical distribution of specific humidity Root Mean Square Error (RMSE) between operational soundings and COSMO-CLM, at three locations within the climate regions, namely, Barcelona (north-eastern Spain; $lon = 2.1°E, lat = 41.4°N$), Nimes (southern France; $lon = 4.4°E, lat = 43.85°N$) and Pratica di Mare (central Italy; $lon = 12.5°E, lat = 41.7°N$). The differences for the RMSE calculation are obtained from every available sounding within the 1-Sep to 20-Nov period.

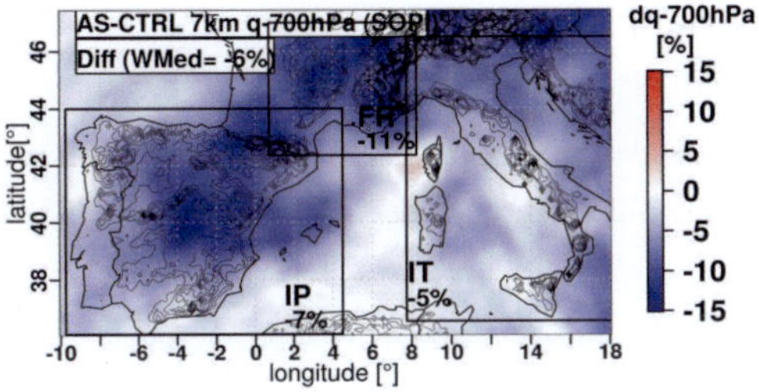

$\frac{AS-CTRL}{CTRL}$ [%]	7 km	2.8 km
IP	-7	-7
FR	-11	-10
IT	-5	-4
VA	-11	-11
BA	-8	-8
CA	-8	-9
CV	-10	-9
CO	-2	-2
CI	-4	-1
NEI	-9	-6
LT	-8	-8

Figure 6.10: Spatial distribution of seasonally averaged specific humidity (q_v) relative differences (AS-CTRL)/CTRL in [%] for the 7 km simulations (left) and table of the relative differences, spatially averaged over the climate regions and the HyMeX target areas (right) at 700 hPa.

and AS-2.8 runs. Below the 950 hPa level, however only the *Pratica di Mare* station shows an improvement in the representation of the specific humidity profile for AS-7 and AS-2.8. At 1000hPa, the Barcelona and Nimes soundings show biases ranging from 1.5 $g\ kg^{-1}$ to 2.4 $g\ kg^{-1}$ for the AS-7 and AS-2.8 simulations, larger by about 0.5 $g\ kg^{-1}$, than their CTRL counterparts. This improvement of the RMSE over 900 and worsening below, stems from the fact that the column moisture nudging reduces humidity in the whole profile. Given the inability of COSMO-CLM to represent a vertical gradient of humidity as strong as that of the soundings in the seasonal values, this reduction is beneficial for the specific humidity representation above 900 hPa. but detrimental to the levels below. This means, COSMO-CLM is underestimating the specific humidity amount above 900 hPa but underestimating it below.

Figure 6.10 shows the differences (AS-CTRL) of the specific humidity at 700 hPa. In agreement with the humidity reduction shown by IWV in section 6.1, specific humidity was reduced at 700 hPa over most of the Mediterranean Sea, Spain, France and Italy. Reductions in humidity of -7% and -7% (IP), of -11% and 10% (FR) and of -5% and -4% (IT), take place for the 7 km and 2.8 km horizontal resolutions, respectively. Again, at 700 hPa, is Italy the climate region less impacted by the column moisture nudging, especially at

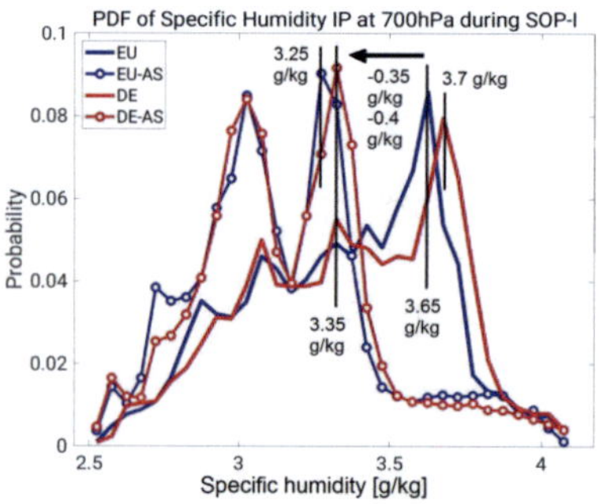

(a) IP (7 km and 2.8 km)

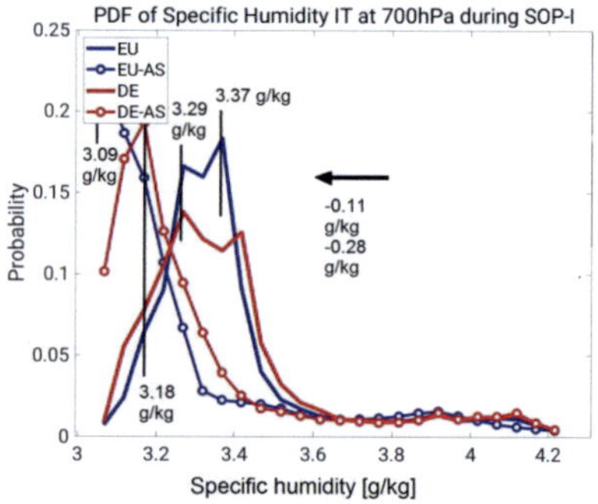

(b) IT (7 km and 2.8 km)

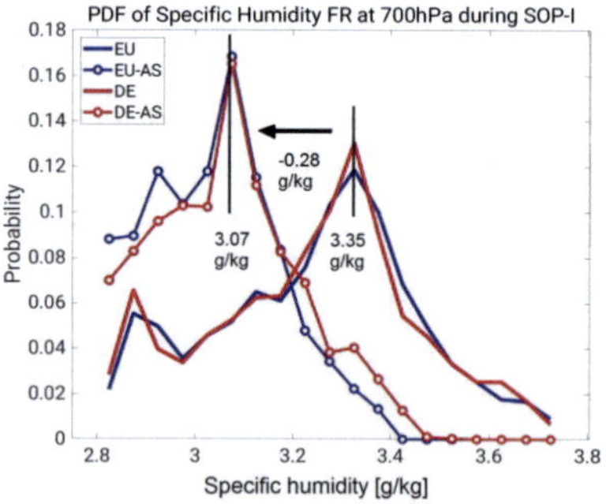

(c) FR (7 km and 2.8 km)

Figure 6.11: Probability Distribution Function (PDF) of the timely averaged (1-Sep to 20-Nov) specific humidity spatial distribution at 700 hPa shown in Fig. 6.10. For all simulations and the climate regions (IP, FR and IT).

the Mediterranean Sea and along the occidental coast. Regarding the smaller HyMex areas, the largest reductions took place over VA and CV, of about -10%, and the smallest over CO and CI, in agreement with the relative differences of IWV.

In order to know the order of variation of the specific humidity spatial distribution after nudging column atmospheric moisture the Probability Distribution Function of the timely means (1-Sep to 20-Nov) of this variable is shown in Figure 6.11 at the 700 hPa level. Figure 6.11, shows how the most probable specific humidity values for CTRL-7 and CTRL-2.8 endure a shift toward lower amounts ranging between $-0.11\ g\ kg^{-1}$ and $-0.4\ g\ kg^{-1}$ in the AS runs, depending on the considered climate regions. This is a very large change, given absolute values over those areas ranged between 2.5 $g\ kg^{-1}$ and 4 $g\ kg^{-1}$ in the seasonal means. It is worth noting the very different PDFs of the CTRL and AS runs for each climate region. IP, shows a predominant value of ca. 3.7 $g\ kg^{-1}$ in the CTRL runs, which is transformed by the nudging into two maxima of

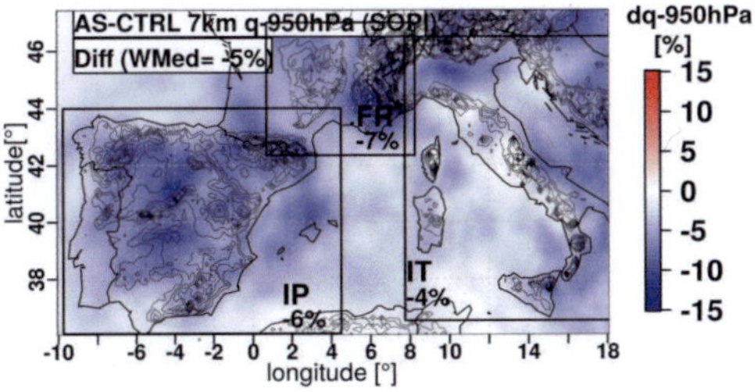

$\frac{AS-CTRL}{CTRL}$ [%]	7 km	2.8 km
IP	-6	-4
FR	-7	-5
IT	-4	-2
VA	-6	-7
BA	-6	-4
CA	-7	-5
CV	-7	-6
CO	-3	-2
CI	-3	-1
NEI	-6	-2
LT	-5	-2

Figure 6.12: As figure 6.10 at for the 950 hPa level

probability in the AS runs (3.3 $g\ kg^{-1}$ and 3 $g\ kg^{-1}$). For their part, IT and FR, only show one clear maximum of probability in the CTRL runs, of about 3.3 $g\ kg^{-1}$, which is shifted toward ca. 3.1 $g\ kg^{-1}$ in the AS runs.

At 850 hPa (not shown), the same overall humidity reduction took place, in agreement with the reduction of specific humidity throughout the atmospheric column. At this level, the relative reductions in the spatially and timely means are of -7% and -7% (IP), of -8% and -5% (FR) and of -5% and -4% (IT), for the 7 km and 2.8 km horizontal resolutions, respectively.

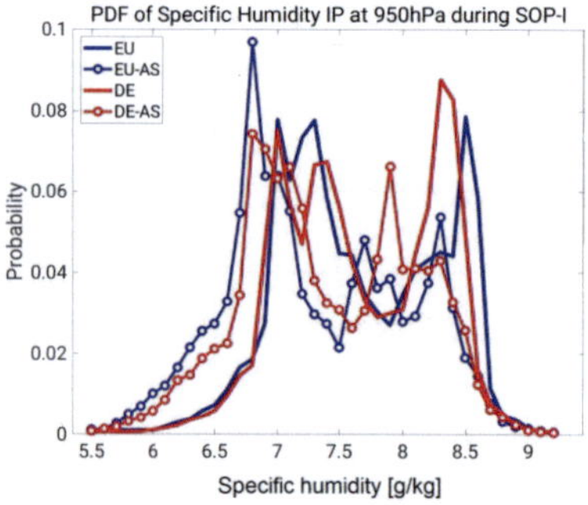

(a) IP

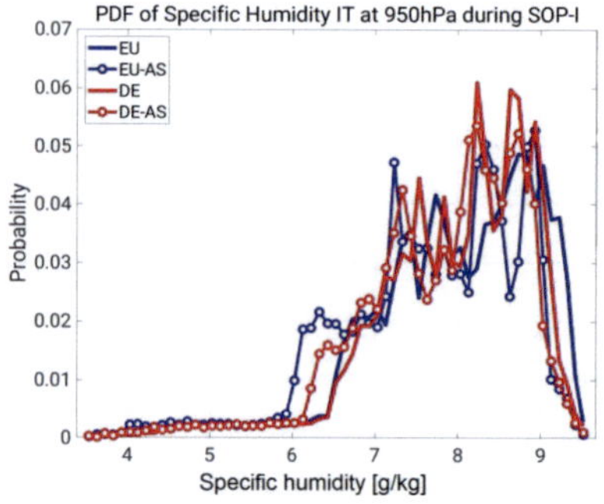

(b) IT

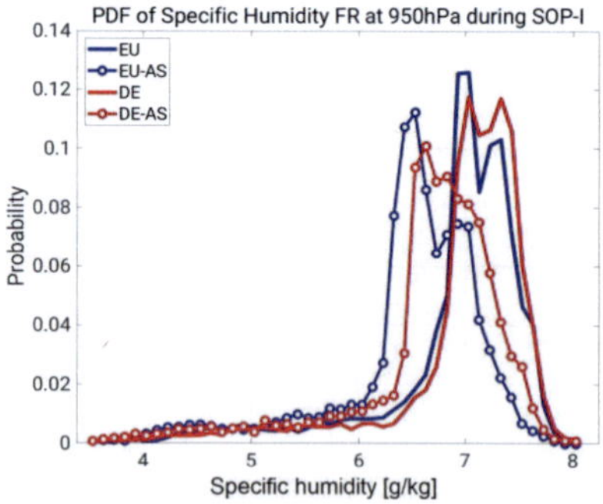

(c) FR)

Figure 6.13: As figure 6.12 at for the 950 hPa level.

At 950 hPa, the relative changes of specific humidity are -6% and -4% (IP), -7% and -5% (FR) and -6% and -4% (IT), for the 7 km and 2.8 km horizontal resolutions, respectively. These relative variations are obtained, in a like manner to those at 700 hPa and are represented in a table in Figure 6.12, together with the spatial distribution for the 7 km grid. Similar percentages of variation are shown for the HyMeX target areas, where, in agreement to IWV and specific humidity at 700 hPa, the largest variations occur over VA and CV, and the shortest variations are found over CI. The order of magnitude of the relative variations at 950 hPa is somewhat smaller, as compared to the 700 hPa level since the specific humidity at this level is larger and thus, for an absolute change of 0.5 $g\ kg^{-1}$, the percentage of relative variations is smaller.
Figure 6.13, represents the PDF of specific humidity at 950 hPa for the timely means (1-Sep to 20-Nov) over the three climate regions. As opposite to the PDFs of specific humidity at 700hPa, there is much more variability in specific humidity at this level. Hence, over IT and IP the most probable values of specific humidity for the CTRL and AS runs of both resolutions are within the range 6.5 $g\ kg^{-1}$ to 8.5 $g\ kg^{-1}$. Over FR this range of variation of the most probable specific humidity value is narrower and is of 6.8 $g\ kg^{-1}$ to 7.5 $g\ kg^{-1}$] for the CTRL runs and of 6.5 $g\ kg^{-1}$ to 7.1 $g\ kg^{-1}$ for the AS runs. In spite of the larger noise in the signal of the PDFs for specific humidity, it is evident the probability reduction of largest specific humidity over the three regions in the AS runs compared to CTRL.

Conclusions

Q.1 *How well is the atmospheric moisture distribution represented in COSMO-CLM for the HyMeX-SOP1, on a seasonal scale? Are there any particular periods or regions of large humidity model biases?*

COSMO-CLM in the near-operational configurations described in Chap. 5 for the 7 km and the 2.8 km grid lengths, shows an overall wet bias over the WMed during the 1-Sep to 20-Nov period (HyMeX-SOP1). This is an issue addressed in past studies using COSMO (Schraff et al., 2008; Devidasrao, 2012; Cress et al., 2012). Differences up to 5 mm exist in the spatial distribution (see Fig. 6.2) and up to 6 mm in the temporal evolution of IWV (see Fig. 6.5). All climate regions show this wet bias. The Root Mean Square Error (RMSE) and Agreement Index (AI) of the forecast runs (CTRL-7 and CTRL-2.8), stayed within the ranges 1.7 mm to 2.1 mm and 0.972 to 0.964 averaged spatially over the broad climate study regions and within the ranges 1.9 mm to 3.7 mm and 0.912 to 0.973 for the local target areas.

One found contributor to this bias is the occurrence of four sub-periods of persistent large IWV biases in COSMO-CLM (up to 6 mm) within SOP1. These are 6-Sep to 12-Sep and 12-Nov to 20-Nov with noticeable biases over all climate regions; 18-Oct to 22-Oct with large biases over FR and 21-Sep to 01-Oct over IT. Daily IWV advection was large during these four periods, as demonstrated by the comparison against the Moderate Resolution Imaging Spectroradiometer (MODIS) data (see Fig. 6.6 and Fig. A.4). The first three periods are, to some extent, influenced by intense evapotranspiration over Spain, northern Africa or Corsica and Sardinia in the COSMO-CLM simulations.

The biases were found to be somewhat larger over IP than over FR or IT in the CTRL runs. Still, after the correction exerted by the column atmospheric moisture nudging it was IT the region showing the largest RMSE. Regarding the smaller target areas, CO and CI showed the worst RMSE and AI values, ca. 3.3 mm and 0.92, respectively in the CTRL runs. These worse values of the verification metrics over CO and CI might be related to the large exposure to the maritime southwesterly flow and proximity of the sea. Besides, Corsica is a region of complex orography, rendering the characterization of IWV difficult and CI shows a GPS coverage gap in the Tuscany and Campania regions.

COSMO-CLM underestimates IWV as compared to GPS, over six particular mountain stations. These overestimations can be as large as 4 mm for the seasonally averaged IWV in the CTRL-7 grid and 3 mm in the CTRL-2.8 grid. This problem stems from the differences between the model and the station height. Using a finer grid resolution (2.8 km) the representation of IWV for

these stations improved, given the more accurate representation of the model surface height.
Regarding the vertical distribution of seasonally averaged specific humidity COSMO-CLM showed Root Mean Square Errors (RMSE) in the order of $2\,g\,kg^{-1}$ between 1000 hPa and 900 hPa for the analysed stations. The comparison against the operational soundings of Barcelona (Spain), Nimes (France) and Pratica di Mare (Italy) revealed an underestimation of humidity below 900 hPa for the two former. Over 900, all three stations show an overestimation of humidity. The reason for this transition from under to overestimation at a particular height stems from the inability of COSMO-CLM to represent a sufficiently strong humidity gradient between the surface and 600 hPa.

Q.2 *What is the impact of nudging column atmospheric moisture on the representation of humidity? How does the continuous GPS-ZTD nudging affect its spatial distribution, the timing, the represented humidity extremes, the moisture transport and the diurnal cycle?*

The column moisture nudging corrects the overall wet bias and improves the mean absolute error of the seasonally averaged spatial distributions (from ca. 1.3 mm to 0.8 mm); and the RMSE (from ca 1.8 mm to ca. 0.5 mm) and the AI (from ca.0.97 to ca. 0.99) of the temporal evolution of IWV. Still, the particular mountain regions with a dry bias persist after the nudging. This indicates that the model-station height difference is too large for the nudging to correct these errors. More specifically, the nudging rejects the assimilation of the ZTD observations measured at those stations since the modelled surface height is more than 100m below the station height. Whence, the nudging procedure avoids extrapolating the input ZTDs to the too low heights. Regarding the improvement over the HyMeX areas, the RMSE and AI values are improved likewise the broader climate regions and lay within the range 0.4 mm to 1 mm for the former and within 0.995 and 0.999 for the latter.
A reduction of IWV in the order of -3% to -8% took place over all climate study regions. Over land points of large GPS coverage the reduction reached larger

values of -10%. The largest reductions took place over the Iberian Peninsula and France, whereas the weakest impact was over IT, specially at the middle and south of the peninsula. A reduction of the same order was also present over the HyMeX target areas, and likewise the broader climate regions, the Italian HyMeX areas, CO and CI were the least impacted by the nudging, about -2%. Over the Sea, in spite the lack of GPS stations, IWV is reduced by about -2% for locations away from the shore. This is due to, first, the spreading of the nudged ZTD information in the horizontal dimension (see Chap. 4 for a detailed description of the weights for horizontal spreading) and second the advection of the re-distributed humidity with the wind flow.
Two characteristics of the representation of IWV which were most affected by the nudging of column atmospheric moisture observations, where the probability for extreme IWV values, which endured a reduction for values above 30 mm, over all domains (see Fig. 6.8) and the intensification of the diurnal cycle. In this regard, the nudging of ZTD was able to reduce vastly IWV in the early morning and more softly at noon (see Fig. 6.7).
The reduction of humidity brought by the nudging was visible throughout the complete atmospheric column as seen in the comparison against operational soundings. This was also assessed, at two significant levels, 700 hPa and 950 hPa showing an overall humidity reduction (over land and sea) varying between -1% and -11%, depending on the considered area, resolution and level. This turned out to be beneficial for the representation of the specific humidity profile above 900 hPa (700 hPa in the case of Pratica di Mare), since the forecast COSMO-CLM runs (CTRL-7 and CTRL-2.8) overestimated humidity above that level. But it was detrimental below. Since humidity was already too low.

Q.3 *What are the differences and similarities of the water vapour representation between the 7 km and the 2.8 km grid settings?*

The following aspects of humidity representation, showed very similar characteristics in the 7 km and 2.8 km runs: the temporal evolution of IWV, the low terrain IWV bias, the vertical distribution of specific humidity and the

Probability Distribution Function (PDF) of the seasonally averaged humidity quantities. These similarities stem from the fact that the runs performed on the 2.8 km grid were forced by the CTRL-7 run (see the used methods in Chap. 5) On the contrary, utilizing a finer grid, showed added value in improving the IWV biases over elevated terrain, given the better modelled surface height and in representing larger extreme IWV values (see Fig. 6.8b);

In the representation of the daily averaged and seasonally averaged IWV, very similar spatial distributions and temporal evolutions were found for IWV and specific humidity between the CTRL -7 and CTRL-2.8 runs and the AS-7 and the AS-2.8 runs, respectively. Only when hourly values were considered, for example in analyzing the PDF of IWV, did the different grid type bring a change.

7 Sensitivity of Convective Processes to Atmospheric Moisture Corrections

Being at the core of the convective processes, atmospheric moisture and its variations have a determinant role in initiating, maintaining and shaping precipitating convection. Chapter 6 showed how nudging column atmospheric moisture brought a large modification of the spatial and temporal distribution of water vapour during the simulation of the HYdrological cycle in the Mediterranean eXperiment (HyMeX)-Special Observation Period 1 (SOP1). The seasonal humidity statistics showed an intense reduction of atmospheric moisture at all pressure levels up to 500 hPa bringing better error and bias metrics in the simulation of Integrated Water vapour (IWV). Given these remarkable changes in the distribution of moisture in the simulation, this chapter aims at assessing the impact on convective precipitation and its preconditioning. The scientific questions addressed in this chapter are:

Q.1 *How well does the Consortium for Small-scale Modelling (COSMO) model represent extreme precipitation during the HyMeX-SOP1? Did the nudging of column atmospheric moisture bring an improvement? How did this impact affect the several features of precipitation on the 7 km and 2.8 km horizontal resolutions? (spatial distribution, temporal evolution, representation of extremes, diurnal cycle, etc.)*

Q.2 *What is the role of the IWV variations preceding heavy precipitation? Are their extent and duration relevant for controlling the intensity of heavy precipitation? Is there a cause-effect relation?*

Q.3 *Which convective processes (and how much) are affected by the moisture nudging on seasonal and event scales? Are any of these, especially relevant to explain the differences in precipitation? If so, is there any tendency for a particular region or period?*

In this chapter, section 7.1 addresses question *Q.1* for the whole SOP1 including validations against remote sensing observations. Section 7.2 presents a novel method for statistical analysis of the IWV-precipitation relationship, quantifying among others the frequency and extent of the IWV increases preceding heavy precipitation. Finally, sections 7.3 and 7.4 investigate the effect of the moisture nudging on the instability, triggering and moisture supply, through changes on convective-related variables (surface temperature and humidity conditions, sensible and latent heat fluxes, convective available potential energy and cloud cover), for the complete season and for selected Heavy Precipitation Events (HPEs) of the period.

7.1 Modelling of Heavy Precipitation during HyMeX-SOP1

In order to quantify the accuracy of the representation of precipitation during the September to November period, the COSMO fields of timely averaged precipitation are compared to a remote sensing precipitation data set, the Climate Prediction Center Morphing Technique (CMORPH) product (see Chap. 5.3 for a description of the data set). This comparison is done by means of an object based verification method, the Structure (S), Amount (A) and Location (L) method. SAL quantifies the accuracy of the precipitation forecasts by means of these three metrics where the best result is indicated by SAL values close to zero (a brief introduction to this method can be found in Chap. 5.1). Figure 7.1, shows the SAL diagrams for the comparison of all simulations against CMORPH over the investigation domains (see Fig. 5.2). The column moisture nudging brings an improvement in the representation of the precipitation structures for both resolutions over all climate regions. Improvement of the Location component is shown by all simulations over the Iberian Peninsula (IP) and Italy (IT) (values in the range 0.09 to 0.11), however, over France (FR), the CTRL runs performed better in setting the location of precipitation as compared to AS. Regarding the precipitation amount, the points in the diagram shift toward negative values after applying the column moisture nudging. This is indicative of the precipitation intensity reduction endured in the AS runs of both resolutions. This turns out to be beneficial for the representation of the structure over France (FR), values close to 0.1, but detrimental over IP and IT.

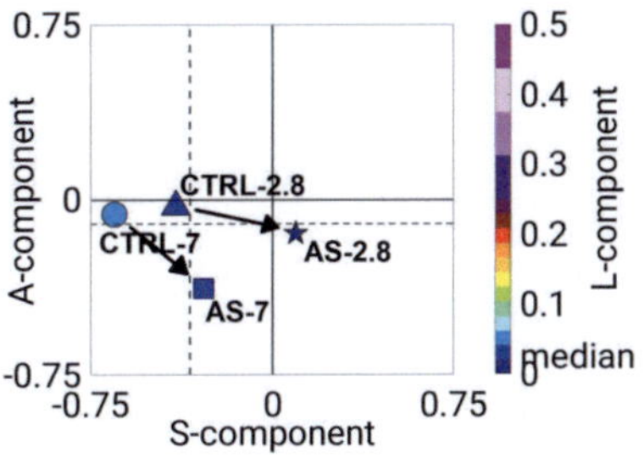

(a) IP

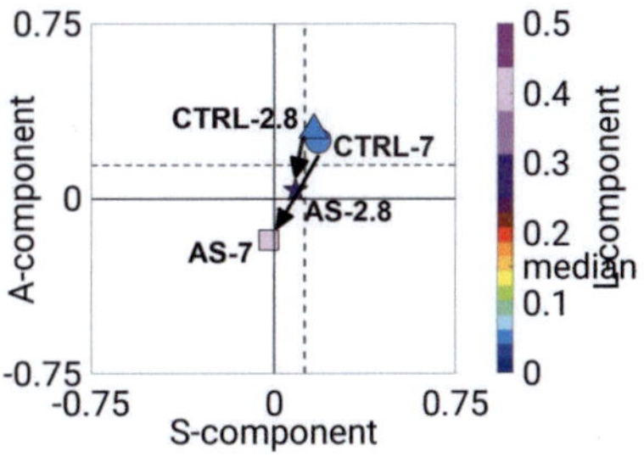

(b) FR

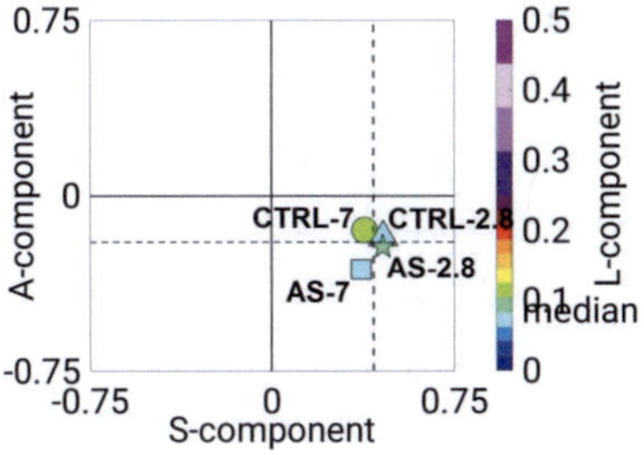

(c) IT

Figure 7.1: SAL verification method applied to timely averaged precipitation (1-Sep to 20-Nov) fields simulated with COSMO and measured by CMORPH. The x-axis stands for the Structure (S) metric, the y-axis, for the Amount (A) component and the colour bar for the Location (L). The closer these metrics to zero, the better the agreement between CMORPH and COSMO.

To further investigate the generalized precipitation decrease as a result of the nudging of column atmospheric moisture, Figure 7.2 shows the seasonally averaged precipitation differences over the Western Mediterranean (WMed) (CTRL-7 and AS-7). Additionally Fig. 7.2 shows the spatially averaged relative differences over the climate regions and HyMeX target areas for both resolutions. All climate regions, show the decrease in precipitation amount of the period, ranging between -9% and -65% depending on the area and horizontal grid considered (see table in Fig. 7.2). FR is the climate region where the strongest decrease takes place (for both grid length types), followed by IP and finally IT (with a weaker reduction in the 2.8 km grid type). These reductions over land points coincide the location of the strongest IWV variations, as shown by Figure 6.3. Considering the local HyMeX areas, the variations in precipitation range from -8% to -93% in the 7 km runs and from -51% to 11%.

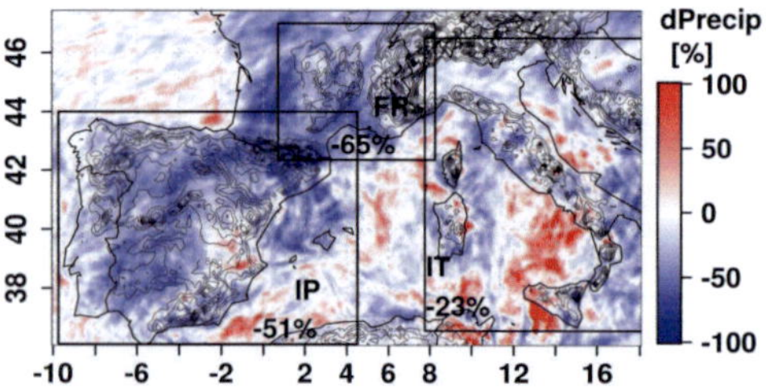

$\frac{AS-CTRL}{CTRL}$ [%]	7 km	2.8 km
IP	-51	-16
IT	-23	-9
FR	-65	-34
VA	-8	1
BA	-40	-11
CA	-93	-51
CV	-65	-6
CO	-26	11
CI	-29	-3
NEI	-27	-10
LT	-28	-4

Figure 7.2: Spatial distribution of the timely averaged precipitation relative differences (AS-CTRL)/CTRL in [%] for the 7 km simulations (left) and table of the percentages of relative differences spatially averaged over the climate regions and the HyMeX target areas.

It is clear thus, that the strong precipitation reduction is also apparent at a local scale. Moreover some regions of about 300km of characteristic length undergo a precipitation increase. This is the case for sea areas such as the Sicilian coast and south to the Iberian peninsula. Concerning the differences due to the use of a different horizontal resolution, the 7 km grid shows larger reductions than the 2.8 km grid. As explained later in section 7.2, the strong moisture reduction causes less triggering of convection in the 7 km runs. This is so since the

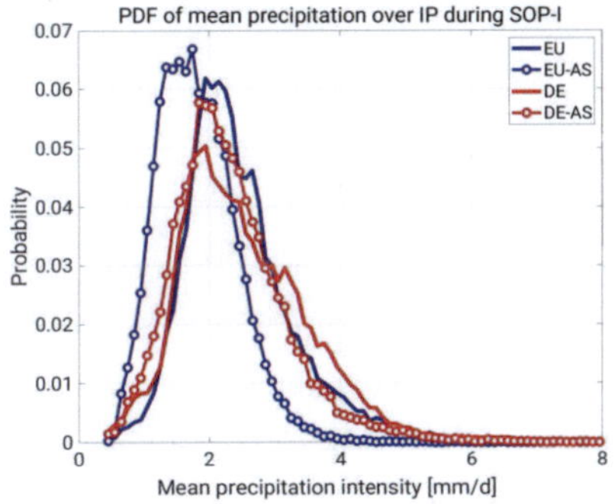

(a) IP (7 km and 2.8 km)

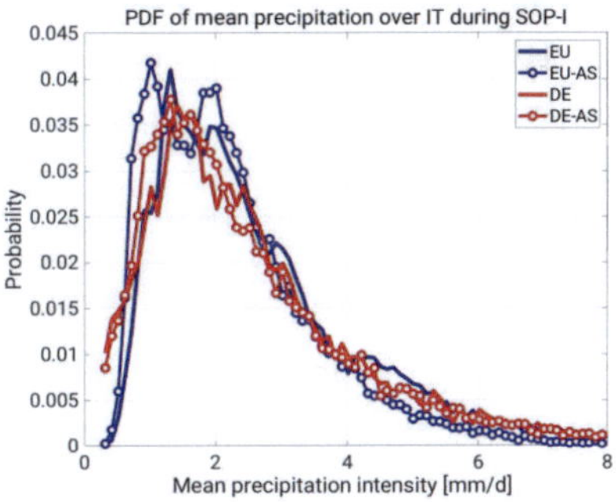

(b) IT (7 km and 2.8 km)

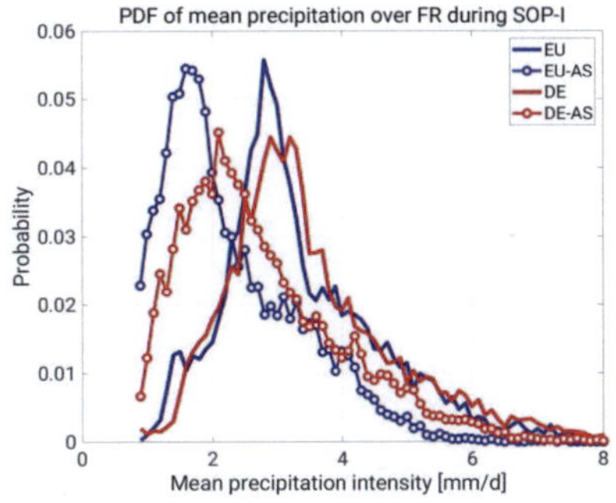

(c) FR (7 km and 2.8 km)

Figure 7.3: Probability Distribution Function (PDF) of the timely averaged (1-Sep to 20-Nov) precipitation spatial distribution over climate regions and for all simulations. The y-axis stands for the probability of a certain precipitation rate to be found in the spatial distribution of the seasonal means.

Tiedtke parameterization scheme (Tiedtke, 1989) imposes a closure condition of sufficient moisture below cloud base for convection to be active.
The generalized precipitation reduction over European land is also perceptible in the Probability Distribution Function (PDF) in the spatial distribution of precipitation shown in Figure 7.3. The column moisture nudging reduces the probability of large values (higher than 4 $mm\ d^{-1}$ in the seasonal averages) for all climate regions and both horizontal resolutions. This shift of 2 $mm\ d^{-1}$ for both resolutions is present over IP and FR. Over IT, however the PDF is much more variable and the impact of the assimilation is less clear.
The precipitation temporal evolution is analysed in the following. Figures 7.4 and 7.5 show daily averaged precipitation, CAPE and hourly values of the Convective Adjustment time-scale (τ), over IP. CAPE is obtained through the mean layer parcel method (see Chap. 2.2). τ gives a quantification of the time scale of CAPE removal by convection, and is useful to classify the degree of forcing exerted by the large scale environment. Following Keil et al. (2008), adequate τ thresholds for differentiating the weakly and the strongly synoptically forced regimes are <3h (strongly forced cases) and >12h for weakly forced cases. The τ scale is a continuum and values within these ranges indicate a combination of the mesoscale factors and the large-scale in forcing the convective events The τ thresholds are shown by the dotted horizontal lines in Figs. 7.4 and 7.5.
The results show that all simulations are able to represent the main heavy precipitation events (as compared to CMORPH observations) therein contained the Intensive Observation Periods (IOPs) of the HyMeX campaign over the eight target areas (see Fig. 3.5). Still, timing lags exist of $\pm 6\ h$, as well as large differences in the precipitation amount. See, for example Fig. 7.4 on the 20-Oct for CTRL-7 or Fig. 7.5 on the 26-Oct for CTRL-2.8.
The simulations showed a large sensitivity to the nudging of humidity data during these periods of heavy rain. The precipitation changes (in relative numbers) induced by the humidity nudging, during the IOPs of the HyMeX period revealed equally increases or reductions, for both grid lengths used (7 km and 2.8 km), albeit the 7 km runs showed a tendency for reduction of precipita-

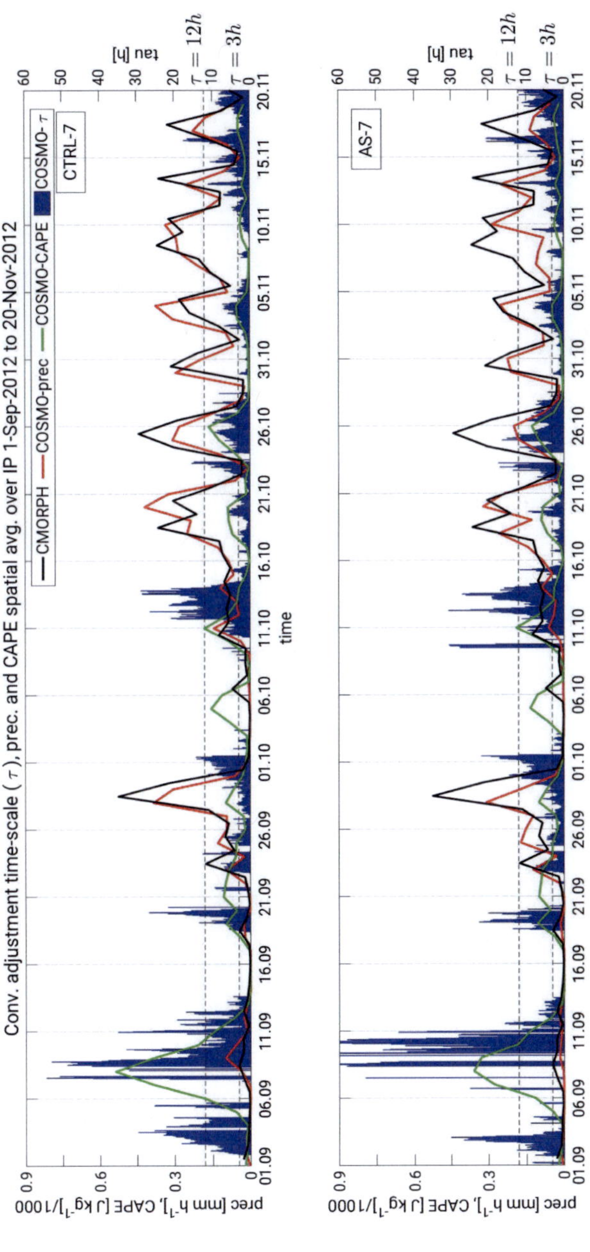

Figure 7.4: Temporal evolution of spatially averaged daily precipitation (COSMO in red and CMORPH in black), CAPE (green) and hourly TAU (blue bars) over IP for CTRL-7 and AS-7, CTRL-2.8 and AS-2.8.

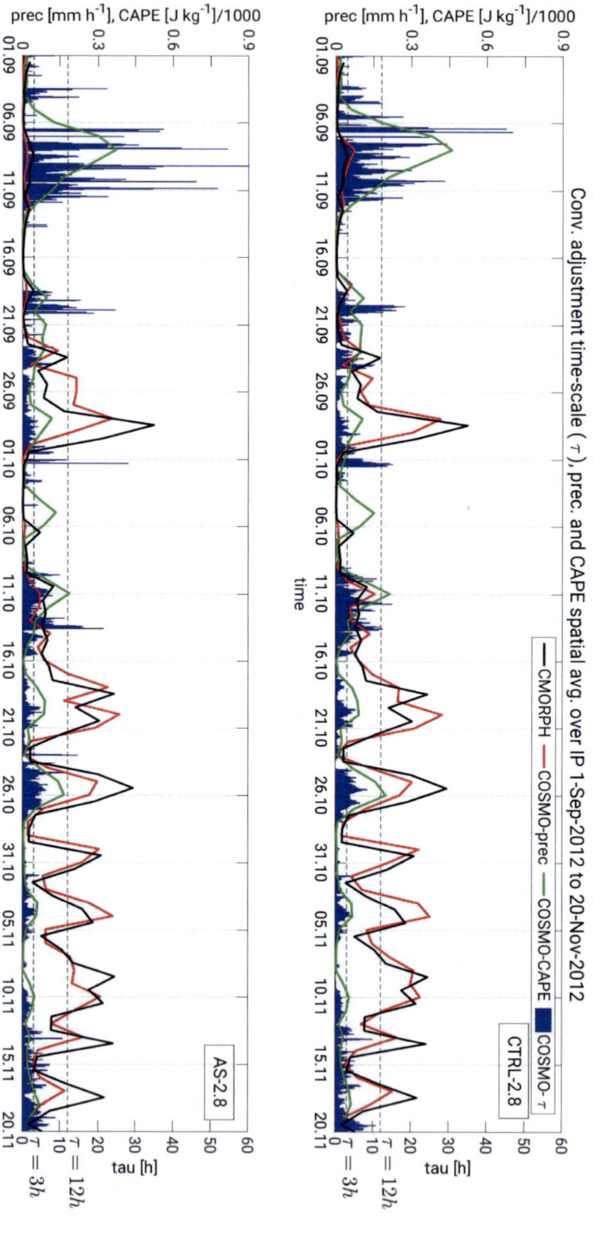

Figure 7.5: As Fig. 7.4 but for CTRL-2.8 and AS-2.8.

tion during the HyMeX IOPs. The largest variations in the 7 km were found over (Catalonia, CA) (IOP18 +100 % and IOP15a -70 %), (Cévennes-Vivarais, CV) (with reductions of ca. -70 % for IOP8 and IOP12a) and Liguria-Tuscany (with reductions of ca. -70 %). For the 2.8 km runs they were found over CA (IOP15a, a weakly forced case with a relative change of -100 %), (Balearic Islands, BA) (IOP18, a strongly forced event with an increase of +300 %), (Corsica, CO) (IOP15c, +140 %) and LT (IOP6 a weakly forced cases and a change of -75 %). A table summarizing the values of the relative variations for each IOP as well as their classification regarding the degree of synoptic forcing is included in Tables A.1 and A.2 of the Annex.

These large increases and reductions in daily precipitation brought only an sporadic improvement in the representation of the HPEs in the period. The fact that the improvement was not systematic highlights the need to take into account the particular features of each event and affected area to understand the underlying reasons for the changes in precipitation. For this reason in section 7.3 an assessment of the impact of the changes exerted by the column moisture nudging on the ingredients of convection is presented for selected HPEs in the period.

Regarding heavy precipitation an analogue decrease of the percentile-90 of the period and of the frequency of heavy precipitation occurs due to the column moisture nudging. Figure 7.6 shows the frequency of precipitation larger than the percentile-90 for CTRL-7, AS-7, CTRL-2.8 and AS-2.8 and for CMORPH in the 1-Sep to 20-Nov period. The CMORPH panel (a) highlights areas prone to heavy precipitation already assessed in past studies of the Mediterranean basin (Lionello, 2012; Ducrocq et al., 2014). These are: The Balkan and Italian occidental coasts, Corsica, the Alpine slopes and foothills, the Pyrenees, the Atlantic Iberian coast, the Atlantic Ocean and the Aquitaine and Poitou-Charentes regions. CMORPH shows, however, a weak performance in the representation of precipitation over the Alps.

It can be observed that CTRL-7 shows an overestimation of the frequency of extreme precipitation over the Iberian Plateau and an underestimation over the

Atlantic Ocean, the Mediterranean Sea and along the Balkan coast. CTRL-2.8 shows the same regions of overestimation and underestimation of probability of heavy precipitation as CTRL-7. However, given the larger values of the percentile-90 precipitation rate (7.2 $mm\ h^{-1}$ for CTRL-2.8 and 6.7 $mm\ h^{-1}$),

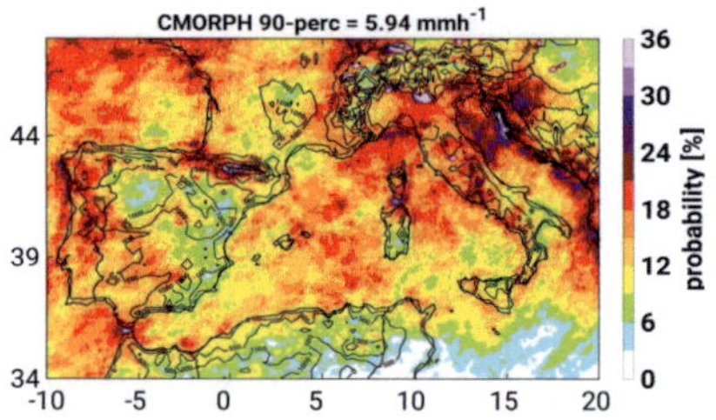

(a) CMORPH

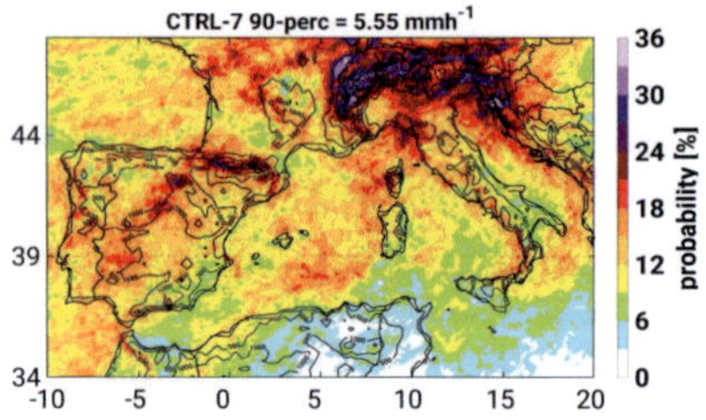

(b) CTRL-7

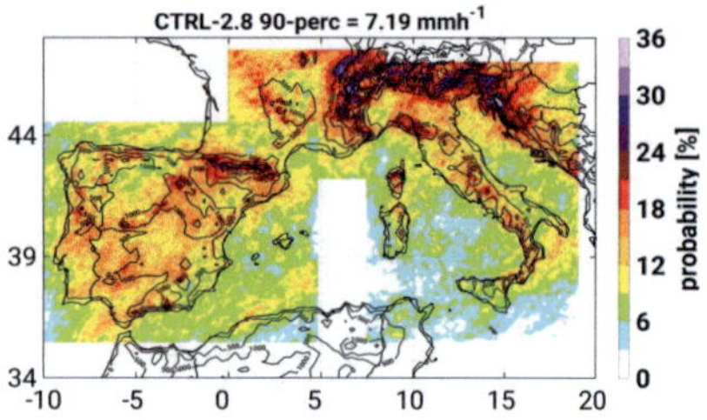

(c) CTRL-2.8

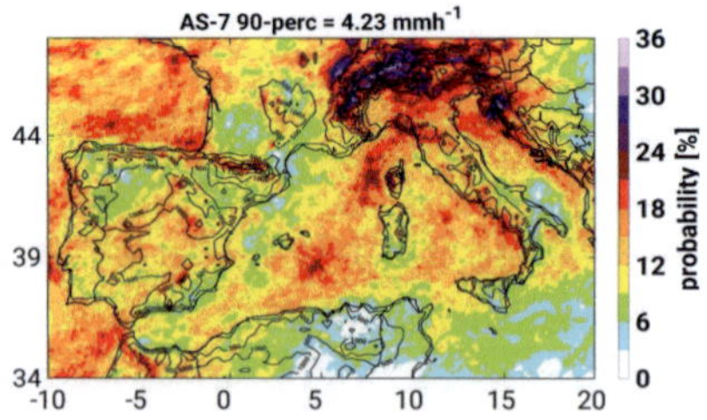

(d) AS-7

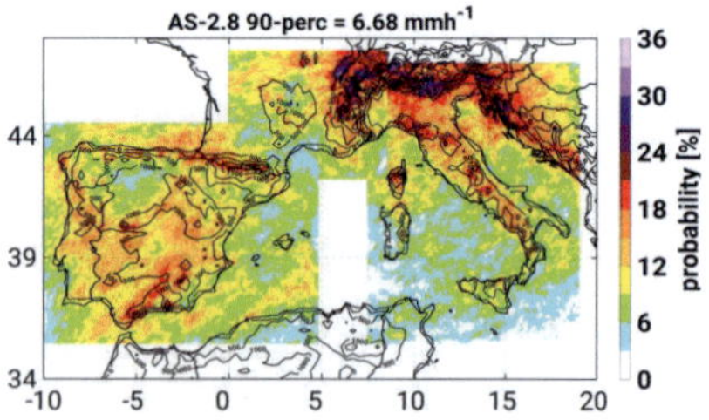

(e) AS-2.8

Figure 7.6: Probability of precipitation larger than the percentile-90 for the 1-Sep to 20-Nov period for CMORPH, CTRL-7, AS-7, CTRL-2.8 and AS-2.8.

the 2.8 km simulations show lower probability than the 7 km runs over the centre of the Iberian Peninsula and along the western Italian coast. Probability values of heavy precipitation drop over these two areas from 20% to 10%, as compared to their 7 km counterparts.

The column moisture nudging works well as to increase the frequency of extreme precipitation over water (Mediterranean and Atlantic), bringing a better agreement with CMORPH, as well as in reducing the frequency over the Iberian Peninsula and France. The nudging of column atmospheric moisture is, however, unable to increase sufficiently the probability of extreme precipitation over the Balkans and both AS-7 and AS-2.8 show probabilities 8 % lower than CMORPH.

How the column moisture nudging impacts the representation of precipitation is further assessed by studying the Probability Distribution Function (PDF) of hourly precipitation for the different climate regions and target areas. In this regard it has been found that applying the nudging in a grid length of 2.8 km, where the Deep Moist Convection (DMC) parameterization can be switched

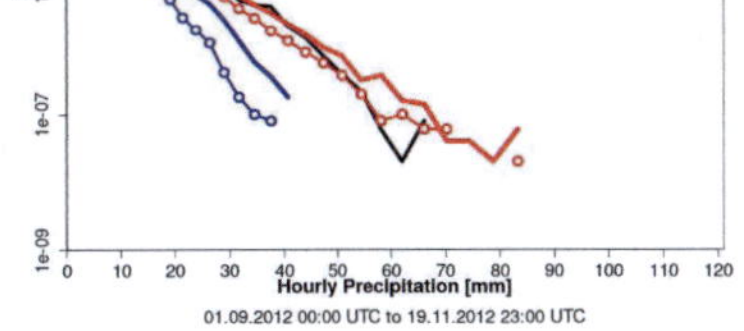

(a) PDF (IP)

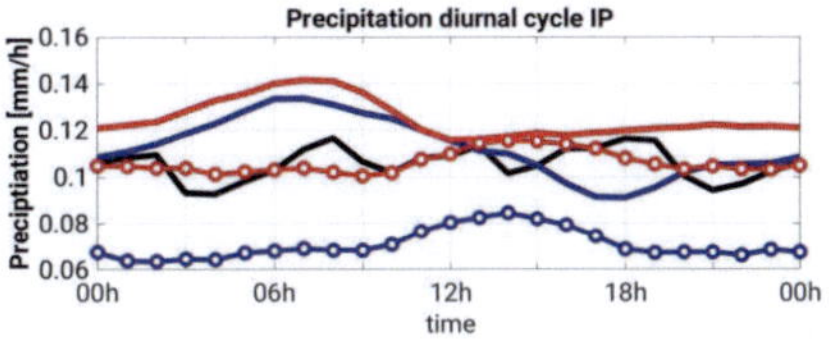

(b) Diurnal Cycle (IP)

Figure 7.7: Probability Distribution Function (PDF) of hourly precipitation from 1-Sep to 20-Nov, spatially averaged for land points over IP and the diurnal cycle. All simulations have been upscaled to the CMORPH grid length of 8km.

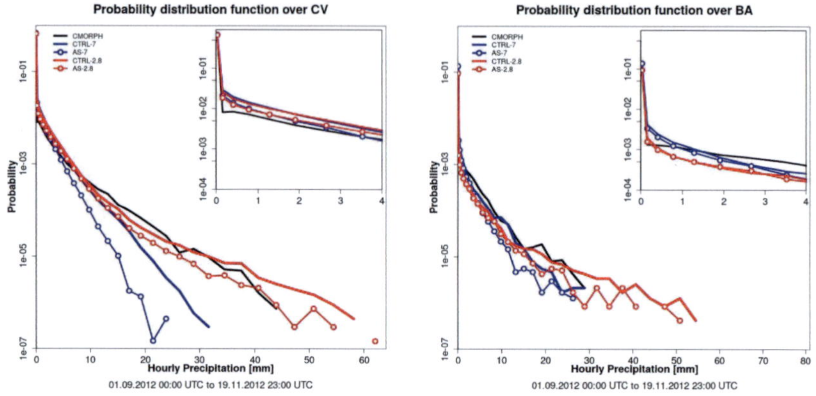

(a) PDF over CV

(b) PDF over BA

Figure 7.8: Probability Distribution Function (PDF) of hourly precipitation over CV and BA for the period 1-Sep to 20-Nov and all simulations. CMORPH observations are represented in Black and only land points are used in the calculation.

off, is beneficial for the representation of the intensity of extreme precipitation. Figure 7.7 shows the PDF of hourly precipitation over the IP climate region. Using a 2.8 km grid increases the probability of precipitation rates above 20 $mm\ d^{-1}$ as well as the maximum precipitation rate (85 $mm\ d^{-1}$). This is also the case over IT and FR (85 $mm\ d^{-1}$ and 60 $mm\ d^{-1}$, respectively). This increased probability of large precipitation can be linked to the use of a model grid resolving explicitly convection as pointed out in past studies (Fosser et al., 2016; Kendon et al., 2017). After nudging the column atmospheric moisture observations the impact for both resolution types and all climate regions is a reduction of the precipitation intensities (see Fig. 7.7). The combination of using a 2.8 km grid length and the moisture nudging brings the best agreement with the observations as measured by the CMORPH gridded product.

The right panel in figure 7.7, shows the diurnal cycle of precipitation over IP for all simulations and the comparison against CMORPH. The nudging of column moisture induced a strong decrease of early morning precipitation and a weaker reduction of precipitation in the afternoon. This, brought a diurnal cycle of precipitation scaling better with the diurnal cycle of solar radiation. The comparison against the CMORPH data set, shows the best results for AS-2.8 over IP and over FR and IT (not shown). This result points out the potential for improvement of the precipitation diurnal cycle representation by combining the nudging of very fast humidity observations into simulations with a fine horizontal grid. In this regard, past modelling experiments (Birch et al., 2014; Fosser et al., 2016; Zhang et al., 2016) have already shown how, alone, reducing the horizontal grid length improves the diurnal cycle of precipitation.

Figure 7.9: Classification of response type to the column moisture nudging in the PDF of hourly precipitation shown by the HyMeX target areas. Five groups of different impacts by the nudging were found depending on the departures between the AS and CTRL runs for precipitation rates larger than 5 $mm\ h^{-1}$.

Prob. of heavy prec. **reduced**	for values [$mm\ h^{-1}$]		BA, CV
	7 km	2.8 km	
	>5	>20	
	for values [$mm\ h^{-1}$]		CA
	7 km	2.8 km	
	>5	>5	
Only 2.8 km prob. of HP **reduced**	for values [$mm\ h^{-1}$]		VA
	7 km	2.8 km	
		>15	
Weak impact	7 km	2.8 km	LT, CI, NEI
Only 2.8 km prob. of HP **increased**	for values [$mm\ h^{-1}$]		CO
	7 km	2.8 km	
		>20	
Only 7 km prob. of HP **reduced**	for values [$mm\ h^{-1}$]		CI
	7 km	2.8 km	
	>10		

Unlike the broader climate regions, the analysis of the moisture nudging impact in the PDF of hourly precipitation over the smaller HyMeX target areas shows a variable impact. Figure 7.8a shows the PDF of hourly precipitation for the 1-Sep to 20-Nov period over the target areas CV (Southern France) and BA (Balearic Islands in Spain). These areas show a reduction of extreme precipitation probability, which takes place as well over VA, due to the moisture column nudging. Different responses to the nudging are observed and the particular features of each region play a decisive role in the representation of precipitation PDF. For moderate precipitation values ($< 5\ mm\ h^{-1}$), the differences between the CTRL and AS runs are negligible over all areas. In contrast, for larger precipitation rates, four responses to the moisture nudging were found in the PDF of hourly precipitation. These are shown in Fig. 7.9, and are a reduction of probability of heavy precipitation in 7 km and 2.8 km (BA, CA and CV; see Figure 7.8), only for 2.8 km (Valencia, VA), only for 7 km (Central Italy, CI) a generally weak impact (North-eastern Italy, NEI and Liguria Tuscany, LT) and an increase of probability of heavy precipitation (CO).

7.2 Integrated Water Vapour Evolution in the Pre-convective Environment

Chapter 6 and section 7.1 have shown that nudging column atmospheric moisture data dries the western European countries, resulting in a general reduction of total precipitation in the simulation of HyMeX-SOP1. These results are evidence of the atmospheric moisture-precipitation relationship which is represented by the model. In spite of being well accepted, the relationship between moisture and precipitation needs further investigation. This section aims at answering the following questions:

- Is there a direct relationship between the increment (reduction) of atmospheric moisture, preceding an increment (reduction) of heavy precipitation?

- What is the relationship between the timing of atmospheric moisture build-up, the extent of the humidity increase and intensity of heavy precipitation?

Identifying IWV changes prior to simulated heavy precipitation during SOP1

The COSMO-CLM seasonal simulations introduced in Table 5.2 are processed to identify the occurring HPEs in the period and the IWV conditions preceding their befallen. This has been done by means of a novel method developed in the course of this thesis, the *IWV-precipitation relationship* method. This method allows to identify the HPEs in the period occurring over the eight HyMeX target areas and the subsequent calculation of statistical quantities. Here, an HPE is defined as a period where spatially averaged precipitation reaches a maximum precipitation of 0.4 $mm\ h^{-1}$ after surpassing a threshold of 0.1 $mm\ h^{-1}$. The event is considered to end when the latter threshold is reached back. Once the events have been identified, the periods of monotonic IWV changes are evaluated. The periods of IWV monotonic changes start at the hour where IWV shifts from increase to reduction or vice versa and the hour of initiation of the corresponding HPE. An explanation of the *IWV-precipitation relationship* method, can be found in Chapter 5. All identified HPEs occurring in the eight HyMeX areas are listed together for each of the four employed COSMO-CLM simulations, in order to reach a sufficient number of HPEs useful for statistical analysis.

Between 124 and 150 events were found to affect the 8 HyMeX areas, depending on the selected simulation (see table 7.1). More events occur in the 2.8 km simulations as compared to the 7 km runs, probably due to the ability of the former to simulate higher precipitation rates. The moisture nudging increased by one event the number of found events in the 7 km runs (AS-7 showed 124 events) and by 14 in the AS-2.8 simulation. The humidity changes induced by

Table 7.1: Number of found events for the different simulations, percentage of events with an IWV increase during the preconditioning and median, mean and standard deviation of mean precipitation. Mean, median and standard deviation of IWV during preconditioning is shown along with the duration of preconditioning.

		CTRL-7	**AS-7**	**CTRL-2.8**	**AS-2.8**
Num. events		124	125	136	150
IWV>0 [%]		85	89	88	75
Precipitation $[mm\ h^{-1}]$	median	0.6	0.5	0.6	0.5
	μ	0.7	0.6	0.7	0.7
	σ	0.4	0.4	0.4	0.4
IWV $[mm]$	median	3.1	2.1	2.9	1.6
	μ	4.6	3.7	4.5	3.1
	σ	5.3	4.4	5	5.1
Precond. duration $[h]$	median	7	5	7	4
	μ	11	8	10	7
	σ	11	8	10	8

the moisture nudging introduced more precipitation variability in the AS-2.8 runs.

Between 75% and 89% of the HPEs, depending on the simulation type, were preceded by an IWV increase, in some cases as high as 25 mm in the spatial averages. The remaining 15% of the cases, showing a decrease in IWV before precipitation onset, took place over areas with a very high moisture content present. This demonstrates that for most cases, IWV monotonic increases precede the precipitating convection in the presented simulations.

To understand how large and for how long the IWV builds up, Table 7.1 provides a quantification of the mean and median values of IWV monotonic changes in the preconditioning period. Mean values between 3.1 mm and 4.6 mm are shown by the different simulations of the set, with the AS-7 and AS-2.8

runs showing smaller values. Again, it is demonstrated that the overall impact of the column moisture nudging was to dry the environment. Besides, Table 7.1 offers a quantification of the mean values of the duration of the preconditioning by moistening of the affected areas in a range of 7 h to 11 h. Again, the as-

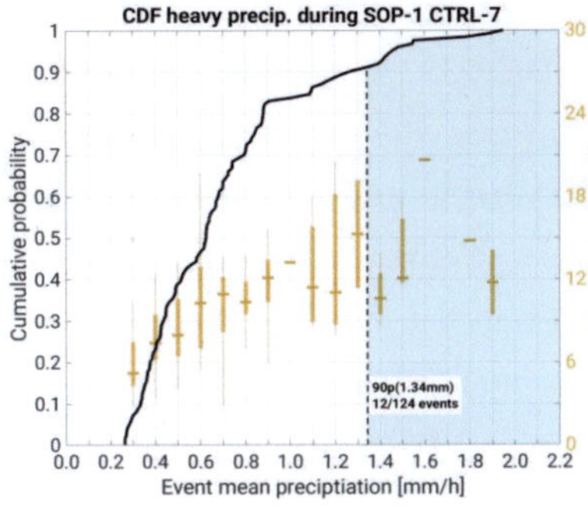

(a) CTRL-7

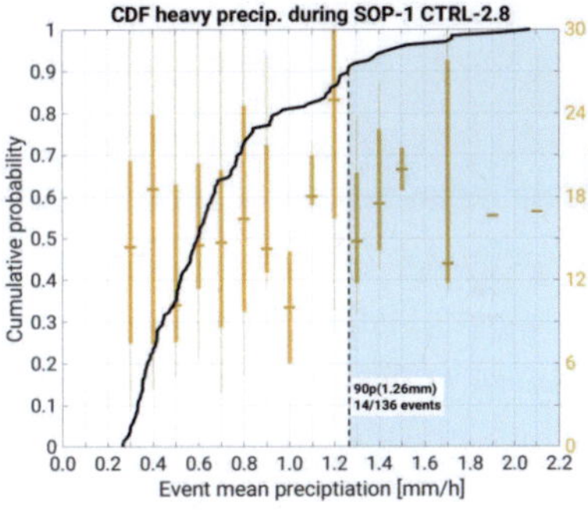

(b) CTRL-2.8

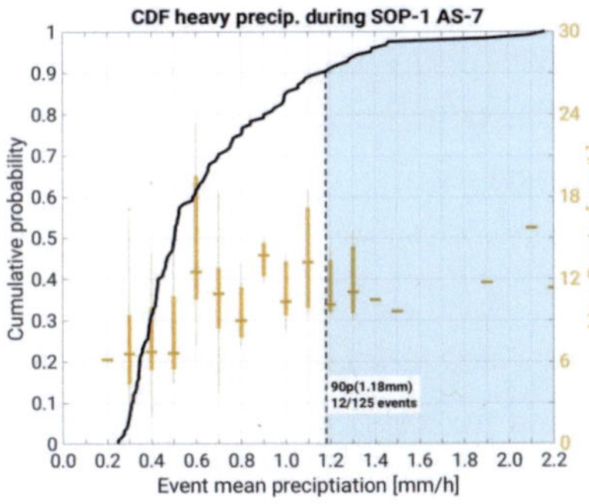

(c) AS-7

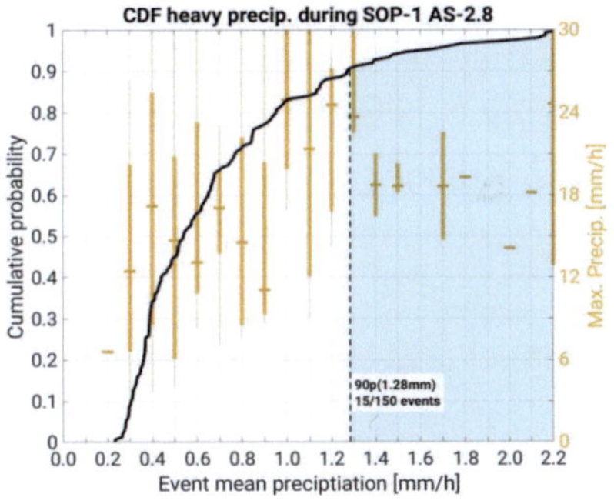

(d) AS-2.8

Figure 7.10: CDFs of timely and spatially averaged hourly precipitation for the identified events in the SOP1 period (black lines). For each simulation all found events are put together to compute the statistical calculations. The left vertical axis represents the cumulative probability, the horizontal axis the mean precipitation of the identified events and the right vertical axes the maximum grid-point precipitation for each of them. The gold box and whiskers, stand for the mean, quartiles and extremes of maximum grid-point precipitation of the events binned in steps of 0.1 $mm\ h^{-1}$ of mean precipitation. the blue shaded contour denotes the percentile-90 similated runs show shorter preconditioning periods than their reference counterparts. This quantification supplements the findings by Schiro et al. (2016), which assessed a maximum humidity amount before the precipitation maxima during the preceding 3h.

In order to know more about the intensity of the identified events in the four simulations, Figure 7.10 shows the Cumulative Distribution Function (CDF) of the timely and spatially averaged precipitation of the found events over the 8 target areas (black line). The left vertical axis shows the probability of a certain mean precipitation value. The right vertical axis shows the maximum grid-point precipitation in $mm\ h^{-1}$ found during each event. The different maximum precipitation values are binned in steps of 0.1 $mm\ h^{-1}$ of mean precipitation in order to know whether the mean precipitation amounts scale with maximum simulated precipitation. A different behaviour of maximum precipitation is found between the 7 km and 2.8 km runs. The former, shows a scaling of maximum precipitation rate with the mean precipitation of the event. Maximum precipitation increases steadily from 2.1 $mm\ h^{-1}$ to 12.5 $mm\ h^{-1}$, for spatially and timely averaged precipitation intensities between 0.2 $mm\ h^{-1}$ and 0.8 $mm\ h^{-1}$. The 2.8 km runs do not show the same behaviour. Even for the events with the lowest precipitation mean amounts, maximum spatially averaged precipitation intensities can be as large as 15$mm\ h^{-1}$. This shows the larger heterogeneity in the spatial distribution of precipitation in a convection permitting grid, where, as reported in past studies, very large precipitation intensities may occur in a very localized area and for a shorter time (Chan et al., 2012; Fosser et al., 2016).

Another relevant aspect is how the duration of the preconditioning, the extent of the IWV increase and mean precipitation relate to each other. Figure 7.11, shows scatter plots relating these three quantities for the four different simulations. Likewise Figure 7.10, every identified event occurring at one of the eight HyMeX target areas is represented (as scattered dots). The horizontal axis represents the mean precipitation of the event, the vertical axis, the monotonic IWV variation preceding the onset of precipitation and the colour stands

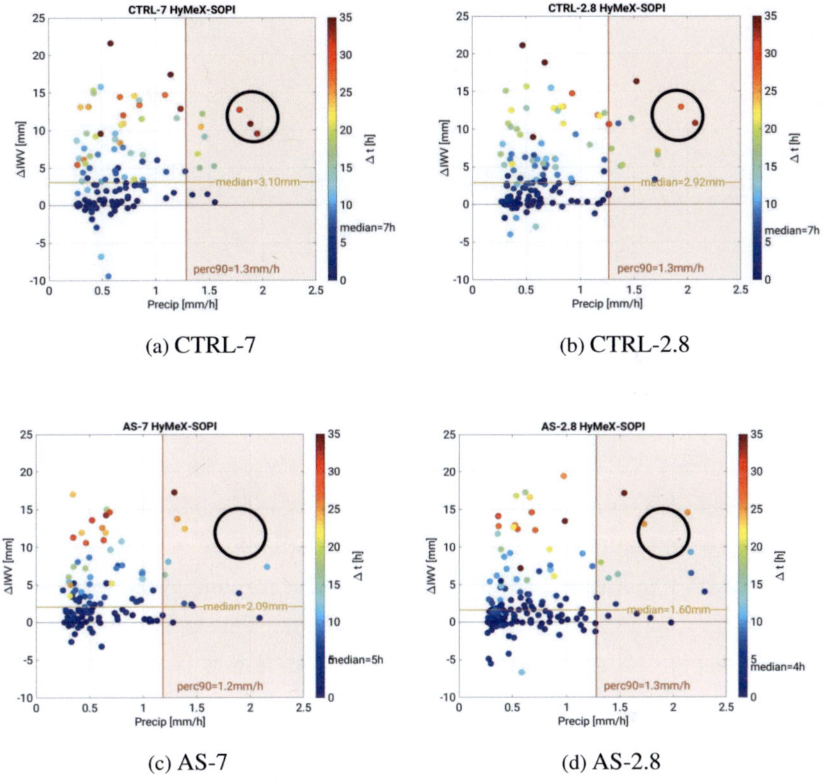

(a) CTRL-7

(b) CTRL-2.8

(c) AS-7

(d) AS-2.8

Figure 7.11: Scatter plots of the extent of IWV monotonic variations (vertical axis), mean precipitation of the events (horizontal axis) and duration of the preconditioning period (colours). Each found event within the simulations, affecting one of the 8 HyMeX target areas is represented by one dot. The median values of ΔIWV and Δt are highlighted as well as the percentile-90 for mean precipitation of the events.

for the duration of the preconditioning period. The top panels, representing the scatter plots of CTRL-7 (a) and CTRL-2.8 (b) show that the events with the largest mean precipitation ($> 1.5\ mm\ h^{-1}$), and denoted by ellipses, are preceded by a very large IWV monotonic increase ($> 10\ mm$) occurring during long preconditioning periods ($\approx 30\ h$). The simulations with nudged Zenith

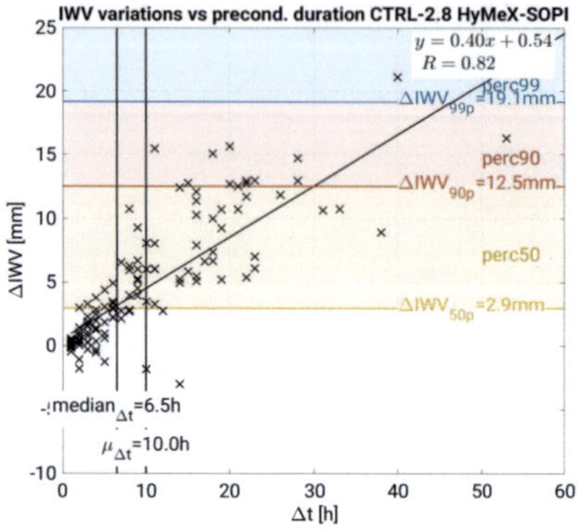

Figure 7.12: Scatterplot of preconditioning duration vs. IWV monotonic variation. Each identified event is represented with crosses. The linear fit only includes IWV increases

Total Delay (ZTD) data, show a different behaviour. Three events, in the AS-7 and AS-2.8 simulations, falling within the percentile-90 of mean precipitation, show low IWV increases ($< 5\ mm$), and very short preconditioning periods (of 1h to 5h), as opposite to their CTRL counterparts. This can be explained by the fact that the AS runs show larger variability of hourly IWV. Therefore the duration of the preconditioning period is shorter, since more easily the IWV temporal evolution changes from a monotonic increase to a reduction or vice versa.

The scatterplots in Figure 7.11, also demonstrate that longer preconditioning periods bring larger IWV monotonic increases. All simulations, show larger duration periods of the preconditioning (reddish colours) for IWV monotonic increases larger than 10 mm. Indeed, it was found that the duration of the moistening is linearly related to the amount of increased IWV where the slope is of 0.4 $mm\ h^{-1}$ (depending on the chosen simulation). Correlation coefficients for this linear relation are within the range 0.81 to 0.86. This result indicates that a particular rate of IWV variation is favoured in the COSMO simulations before the initiation of heavy precipitation. An example is shown for the CTRL-2.8 simulation in Figure 7.12, but this also holds for CTRL-7, AS-7 and AS-2.8.

This IWV variation *speed*, can be related to the changes of the moisture budget equation (Lamb et al., 2012; González-Rojí et al., 2018). Provided the method avoids selecting preconditioning periods with precipitation present, the driver of this favoured IWV *speed* is the moisture advection.

Precipitation changes induced by the nudging of column atmospheric moisture

Four main findings have been stressed out from the previous section.

- An IWV increase precedes heavy precipitation in ca. 90% of the events simulated during SOP1
- In mean, the moistening of the target areas affected by heavy precipitation was of about 4mm (IWV) and lasted for 9h
- The heaviest precipitation events had the longest preconditioning periods ($\approx 30h$) and largest IWV increases ($> 10mm$).
- The duration of the moistening of the target areas showed a linear relationship (correlation coefficients between 0.81 and 0.86) with the extent of the IWV increases with a slope of 0.4 $mm\ h^{-1}$

To delve further into the relationship between moisture build-ups and heavy precipitation, the COSMO-CLM simulations of SOP1 are used once more in the following analysis. The identified events affecting the eight HyMeX areas are categorized further based upon the identification of the HPEs of the period, by means of the *IWV-precipitation relationship* method. The events are now classified depending on whether the column moisture nudging reduced or increased IWV and precipitation. For each grid length type (7 km and 2.8 km) all identified events of the CTRL and AS simulations are investigated to find events starting within a $\pm 6h$ time window. The pairs of events found in the CTRL and their AS counterparts would then be considered as the "same" event. This is acceptable since most events of the period occur equally on the CTRL and AS runs usually within a time window of 12h. Figure 7.13 shows a

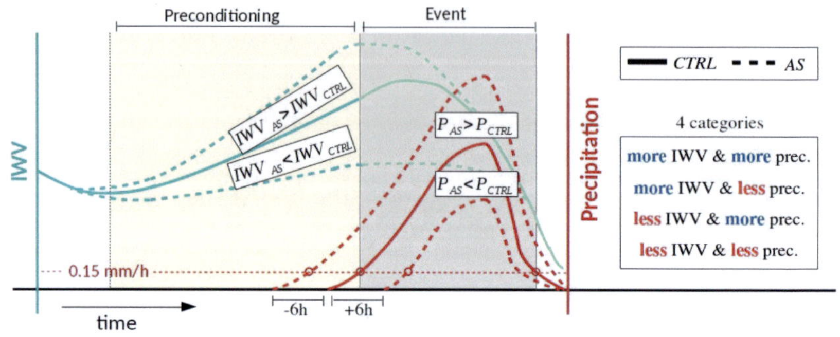

Figure 7.13: Concept of selection of "simultaneous" events in the CTRL and AS runs. The precipitation periods as well as the preconditioning periods are selected using the same criteria as in the *IWV-precipitation relationship* method, described in Chap. 5. Two events simulated by the CTRL and AS runs are considered simultaneous if they occur within a $\pm 6\ h$ time-window. Afterwards, the mean IWV of the preconditioning period and the mean precipitation of the event period are compared to classify the events into four categories. These categories result from the four possible combinations of the IWV and precipitation of the AS runs being larger (shorter) than that of CTRL.

schematic representation of the selection of coinciding events, where straight lines represent the temporal evolution of IWV (blue) and precipitation (red) of CTRL and dotted lines stand for the temporal evolution of the AS runs. Applying the same requirements for event selection (maximum precipitation should reach 0.4 $mm\ h^{-1}$, after surpassing the 0.1 $mm\ h^{-1}$ threshold) and for the preconditioning period (time between the last change of IWV variation sign and the start of precipitation) a total of 80 coinciding events were found in the 7 km and 104 in the 2.8 km runs (see Tables 7.2 and 7.3). Through this approach, more understanding can be gained on how often an IWV increase (decrease) induced by the column moisture nudging brings an increase (decrease) in precipitation. Hence, four categories are possible (summarized in Fig. 7.13)
Tables 7.2 and 7.3 show the percentage of simultaneous events belonging to the four categories (first column). For example, Table 7.3 shows that that 29 % of the events occurring in both the CTRL-2.8 and AS-2.8 runs underwent an IWV

Table 7.2: Percentages of coinciding events in CTRL-7 and AS-7 with the four combinations of IWV and precipitation increases and decreases for the 7 km runs. The identified events are further categorized depending on the degree of synoptic forcing.

7 km	Coinc. events	All events [%]	Strongly [%]	Weakly [%]
more IWV more Precip.	12	15	17	12
more IWV less Precip.	15	19	19	18
less IWV more Precip.	20	25	24	27
less IWV less Precip.	33	41	40	43

decrease as a result of the column atmospheric moisture nudging and a subsequent precipitation reduction. The rightmost columns account for the same categories but after splitting the events depending on the degree of synoptic forcing in the CTRL runs (strongly and weakly, see Chap 5 for an explanation of the categorization).

One of the first results of this analysis, is the fact that for both horizontal resolution types, ca. 60 % of the simultaneous events showed a precipitation change of the same sign as that of the changes of IWV during the preconditioning. This holds for the simulations using both grid length types. In other words, that for events where the nudging of column atmospheric moisture dried (moistened) the environment prior to precipitation in 60 % of the cases precipitation was reduced (increased).

A different behaviour is shown by the 7 km and the 2.8 km runs in the sense that, out of the 60 % of the cases with IWV-precipitation changes of the same

Table 7.3: As table 7.2 for the coinciding events of the 2.8 km runs

2.8 km	Coinc. events	All events [%]	Strongly [%]	Weakly [%]
more IWV more Precip.	31	30	31	25
more IWV less Precip.	21	20	18	26
less IWV more Precip.	23	22	21	26
less IWV less Precip.	29	28	30	23

sign, the 7 km grid length favours those with less IWV in the preconditioning followed by less mean precipitation (in 41 % of cases). On the contrary, the 2.8 km runs do not favour this situation and, indeed, the cases with less IWV and less precipitation in the AS-2.8 run show a very similar percentage (29 %) to those with more IWV and more precipitation (31 %). When split into the different categories of degree of synoptic forcing (strongly and weakly) this behaviour in the 7 km is clearer. Out of the 33 HPEs characterized as weakly synoptically forced cases, 14 (43 %) show less IWV in the AS-7 than CTRL-7 and precipitation is reduced. On the other hand, weak synoptically forced cases in the 2.8 km grid still show equal probability of either case of the same sign (23% in a decrease-IWV, decrease-Precipitation and 25% in an increase-IWV increase-Precipitation situation).

A plausible explanation for this finding relies on the use of the Tiedke convection parameterization scheme in the 7 km runs. The Tiedtke deep convection scheme (Tiedtke, 1989) is of the low-level control scheme type (see Chap. 4), which are very dependent on below cloud base moisture to decide whether convection will be triggered or not. In the Tiedtke scheme the equation controlling whether convection triggers or not, is the budget equation for humidity below the cloud base (Doms et al., 2011).

$$\{M_u(q_u - q) + M_d(q_d - q)\}_{z_b} = -\int_{z_s}^{z_b} \left(\rho \boldsymbol{v} \cdot \nabla(q) + \frac{\partial F^q}{\partial z}\right) dz \qquad (7.1)$$

Where z_s is terrain height, z_b the height of the cloud base, F^q is the vertical turbulent flux of specific humidity, q is the specific humidity, M_u and M_d are the updraft and downdraft mass fluxes, respectively. The second term in brackets of the right-hand side of the equation, controlling the vertical turbulent flux of specific humidity is much more sensitive to the fast humidity changes induced by the column moisture observations. Convection will only occur when the right hand side of Eq. 7.1 is positive, i.e. when moisture convergence tends to increase the sub-cloud moisture (Doms et al., 2011). This explains why the less IWV, less precipitation case is favoured for the 7 km resolution, since this closure assumption sets a requirement in which under insufficient moisture convergence, convection will not be triggered. Situations of weak synoptic forcing are more sensitive to the humidity changes induced by the moisture nudging, that is why the percentage of less-IWV and less-precipitation situation is favoured in the weakly synoptically forced events. For its part, the 2.8m resolution simulations do not use a parameterization scheme for DMC triggering and the situation of less-IWV and less precipitation are equally probable, in situations of strongly and weakly forced synoptic situations, and very similar to the situation of more-IWV vs. more-precipitation

7.3 Sensitivity of the Seasonal Convective-related Variables

Sections 7.1 and 7.2 have demonstrated the large impact of the column atmospheric moisture nudging in heavy precipitation modelling during the SOP1. Among the assessed effects, it stands out the overall decrease in precipitation amount and frequency of heavy precipitation. The frequency of extreme precipitation showed a reduction in probability for the assimilated runs over all climate regions (see figure 7.6). This impact showed a different behaviour at the local scale. When considering the local HyMeX target areas, diverse responses were observed depending on the considered area. For example, reduction of the probability of heavy precipitation over BA and CV, or increases over CO. As pointed out in the conclusions of section 7.1 this dependency of the impact on the characteristics of the affected regions (climate regimes, soil conditions, orography) should be considered. This section aims at understanding how, and how much, the column atmospheric moisture nudging affects the atmospheric variables related to convective processes, taking into account the particular characteristics of each studied region and the particular implications of using a 7 km or a 2.8 km model configuration.

Sensible and Latent Heat fluxes at the surface

The exchange of heat and moisture in the ground-atmosphere boundary is affected by the column moisture nudging, directly, through the instant changes of humidity but also, indirectly at longer temporal scales. Provided the simulations run for three months, the changes in humidity affect the fluxes at the surface by, among others, precipitation, radiation (given the modified cloud cover) and atmospheric moisture advection.
The amount of sensible/latent heat flux absorbed or released at the surface-atmosphere boundary depends on the difference of humidity and temperature between the bottom atmospheric model layer and the soil surface (see Chap. 4.1). Where diffusion of heat or moisture occurs in the direction of the least humid/warm system i.e. the soil or the atmosphere. Table 7.4, shows the spatial distribution of the season averaged 2m-q_v and outbound surface Latent Heat (LH) flux differences between AS-7 and CTRL-7 and AS-2.8 and CTRL-2.8 for the three study climate regions. This representation is analogous to that of Fig. 7.2 for precipitation. All climate regions, for both grid length types, show

a strong reduction of humidity at the surface (of about 0.4 $g\ kg^{-1}$ in mean). The impact, is larger for FR and IP (changes of ca 0.4 $g\ kg^{-1}$) than for IT, given the stronger IWV reduction over IP and FR due to the larger coverage

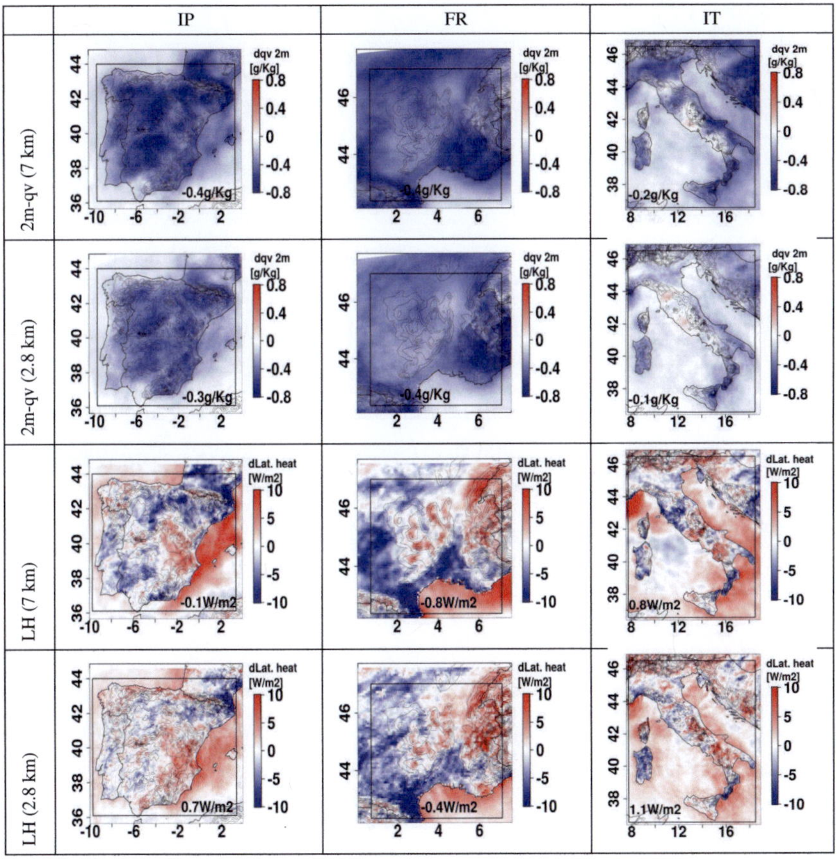

Table 7.4: Spatial distribution of the seasonally averaged differences (AS-CTRL) of specific humidity (q_v) at 2m height in [$g\ kg^{-1}$] and surface latent heat (LH) outbound flux in [$W\ m^{-2}$]. The differences between the seasonally averaged values of AS and CTRL are shown for the two employed grid lengths (7 km and 2.8 km) and the three investigated climate regions. Red colours account for an increase of the quantities after nudging column atmospheric moisture and blue for a decrease.

Table 7.5: Timely averaged differences (1-Sep to 20-Nov) of specific humidity at 2m and surface outbound Latent Heat flux, spatially averaged for the climate regions (IP, FR and IT) and for both grid length types. The differences are obtained in absolute values ($AS - CTRL$).

2m-q_v ($g\ kg^{-1}$)	IP	FR	IT
7 km	-0.4	-0.4	-0.2
2.8 km	-0.3	-0.4	-0.1

Lat. Heat flux ($W\ m^{-2}$)	IP	FR	IT
7 km	-0.1	-0.8	0.8
2.8 km	0.7	-0.4	1.1

of stations (see Fig. 6.3). The Sea, is also largely affected in spite of having no GPS coverage, specially east to the eastern Spanish coast and the gulf of Lions as well as east to Corsica and Sardinia and over the Adriatic Sea. As explained in Section 6.1 the information of the nudged ZTD observations is spread horizontally (see Eq. 4.7) and transported with the wind flow.

Regarding LH flux at the surface, the impact of the moisture nudging is much less homogeneous and the spatial distribution shows patches of increases and reductions equally for both grid length types. The absolute changes in seasonally averaged outbound LH flux can be as high as 10 $W\ m^{-2}$, and show a large variability in the spatial distribution. Three factors explain this heterogeneous spatial distribution. First, moisture reservoirs of the atmosphere (e.g. water parts and very high mountains), emitted more LH flux in the AS runs, given the nudging of column moisture dried the atmosphere close to the surface (as shown by the spatial distribution of the $2m - q_v$ differences). Second, there are some regions such as the Thyrrenean Sea or the Guadalquivir Valley less affected in general by the moisture nudging. Possibly because the moisture advection to these regions, usually from the southwest, hover locations of no GPS coverage. Finally, some land parts of elevations below 3000m underwent an intensification of the LH flux emission (eastern Iberian Plateau, northern Rhone Valley or parts of the Italian Peninsula) in the order of $+8\ W\ m^{-2}$, whereas other (southern France or north-eastern Spain) showed less outbound LH flux in the AS runs. Their spatial distribution matches roughly that of the differ-

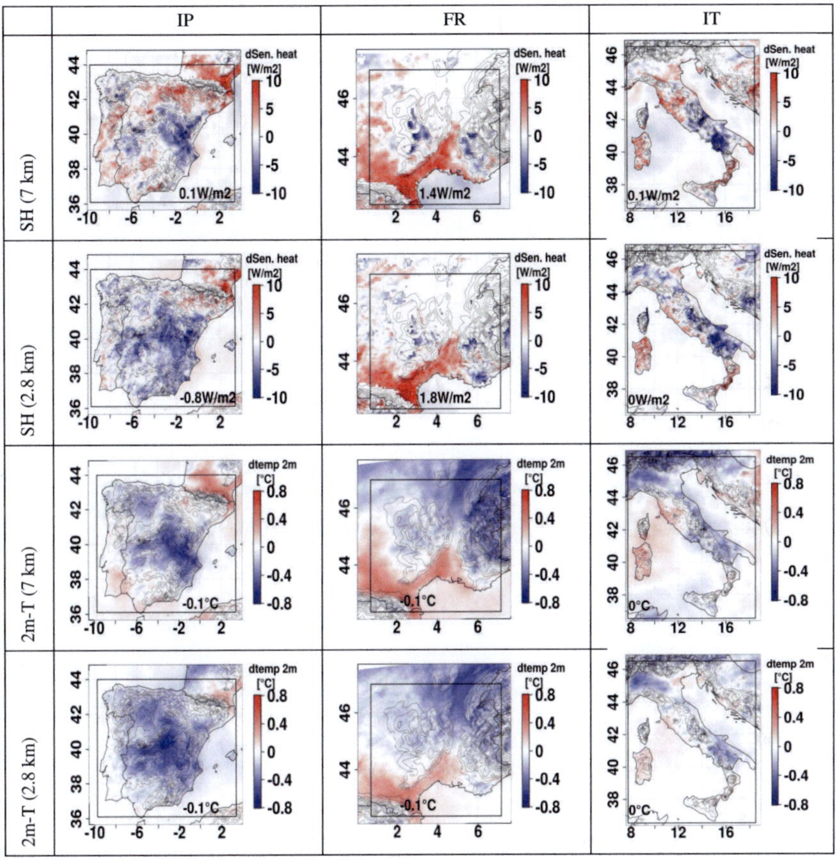

Table 7.6: As Tab. 7.4 for surface Sensible Heat (SH) outbound flux in $[W\ m^{-2}]$ and Temperature at the 2m height in $[°C]$.

ences in precipitation. In other words, in the case of a region showing more seasonally averaged precipitation, more LH is emitted in average in the season. This result holds for the simulations on the 7 km and 2.8 km grid lengths.

As shown in section 2.2 the net radiation at the ground-atmosphere surface is partitioned into latent and sensible heat fluxes (SH). Hence, there where the

Table 7.7: As in Tab. 7.5 but for surface outbound Sensible Heat flux and 2m-temperature.

Sen. Heat flux ($W\ m^{-2}$)	IP	FR	IT	2m-temp ($°C$)	IP	FR	IT
7 km	0.1	1.4	0.1	7 km	-0.1	-0.1	0
2.8 km	-0.8	1.8	0	2.8 km	-0.1	-0.1	0

column atmospheric moisture induced an increase (reduction) of the outbound LH flux, SH flux shows the opposite sign of the variations. Regions showing a reduced emission of SH flux in the AS runs in both resolutions are the Alps, the Pyrenees the Massif Central, the Apennines, coastal regions and the south-eastern Iberian Plateau. These are regions where outbound LH flux was intensified. Past studies have already assessed this effect in COSMO, however on sensitivity experiments modifying the amount of soil moisture (Barthlott and Kalthoff, 2011; Gantner et al., 2017).

Given these variations in the seasonal conditions of SH flux, temperature at the surface is also largely affected. In this regard, a large agreement between the spatial distribution of outbound SH flux and 2m-temperature exists over the climate regions for both grid length types. Figure 7.6 shows the spatial distribution of the season averaged 2m-temperature differences between AS-7 and CTRL-7 and AS-2.8 and CTRL-2.8. The largest temperature differences were of the order of 1 °. After spatially averaging the values, due to the large variability of the spatial distribution, the differences remained small, see Table 7.7.

The spatial distribution of the differences on both the 7 km and 2.8 km simulations showed very similar regions of cooling and warming, albeit some discrepancies over IP at the southwestern coast and over IT between Sardinia and the Italian Peninsula. The reason behind is the smaller precipitation differences in the 2.8 km runs over IP and IT, which in turn induce weaker variations of latent and SH fluxes.

Atmospheric instability

As a measure of the impact of the column atmospheric moisture nudging on the spatial distribution of atmospheric instability, the differences in CAPE between AS-7 and CTRL-7 and between AS-2.8 and CTRL-2.8 are presented in Table 7.9. Analogously to the analysis of the 2m-temperature, the values for the spatial averages are presented in table 7.8. The results show a clear reduction of latent instability as given by CAPE of the mean layer parcel. See Chap. 2.2 for a description of the method.
IP and IT show decreases of spatially averaged CAP in the order of $-10\,J\,kg^{-1}$ and $-20\,J\,kg^{-1}$, respectively. FR shows weaker reductions of $-1.1\,J\,kg^{-1}$ in the AS-7 runs and of $-5\,J\,kg^{-1}$ in the AS-2.8 simulation (see Tab. 7.8). These values are low compared with values reported by past heavy precipitation studies. This is so since they account for a temporal and a spatial average between 1-Sep to 20-Nov for the mentioned areas. The spatial distribution of the differences in CAPE, shows large reductions over the Sea (about $-75\,J\,kg^{-1}$ in the temporal means), specially close to the coasts of eastern Spanish or southern Italy. Over land, the reduction of CAPE for both grid length types is also remarkable, with reductions up to $-50\,J\,kg^{-1}$ in the seasonal averages.
Small differences between the spatial distribution of CAPE variations between the 7 km and the 2.8 km are visible over the Massif Central, the Alps, the Gulf of Liguria and the Adriatic Sea with larger enhancement of CAPE values in the AS-2.8 runs. These different spatial distribution between both horizontal

Table 7.8: As in Tab. 7.5 but for CAPE as obtained by the mean layer method and total cloud cover

CAPE ($J\,kg^{-1}$)	IP	FR	IT
7 km	-10.1	-1.1	-21.3
2.8 km	-10.2	-4.5	-19.9

Total Cloud Cover (%)	IP	FR	IT
7 km	-3.6	-4.3	-2.2
2.8 km	-3.3	-3.8	-1.2

	IP	FR	IT
CAPE (7 km)	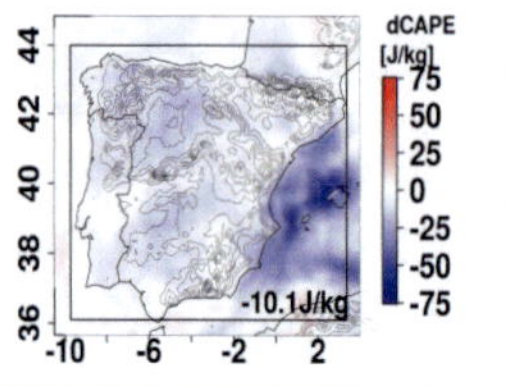	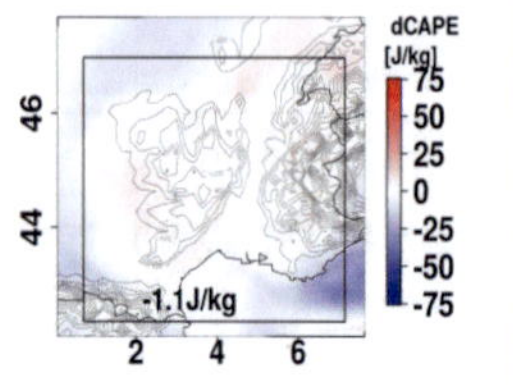	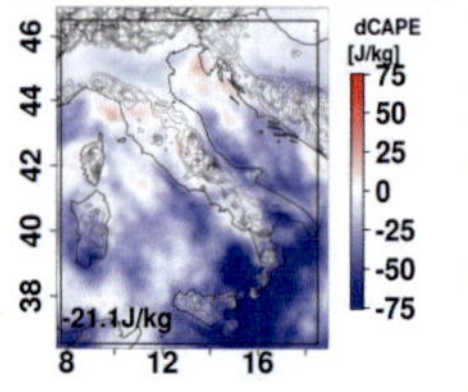
CAPE (2.8 km)	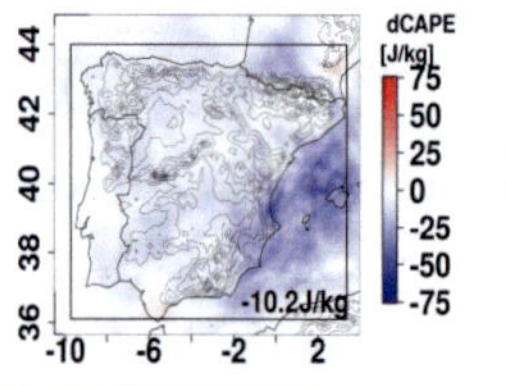	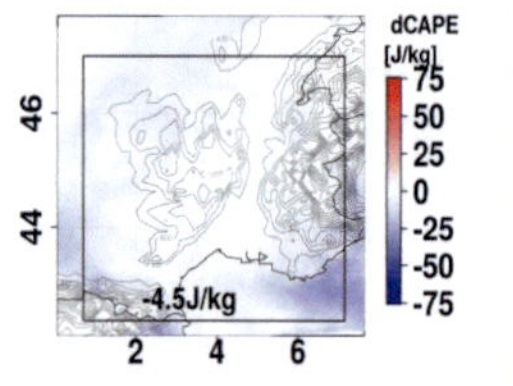	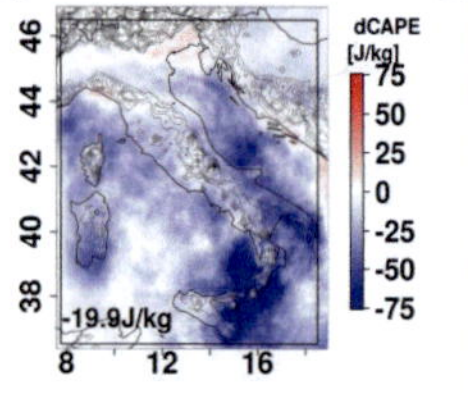
CLCT (7 km)	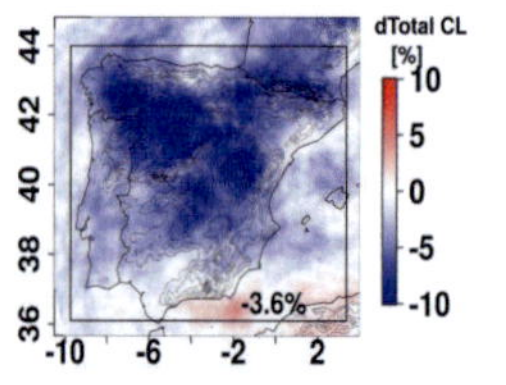	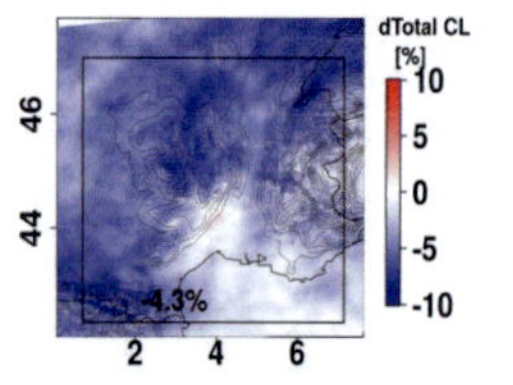	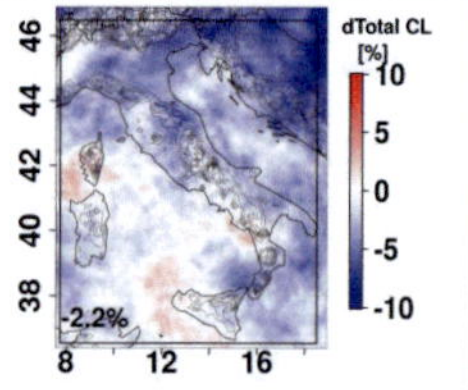
CLCT (2.8 km)	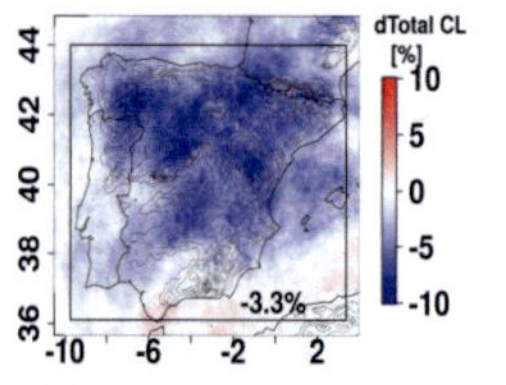	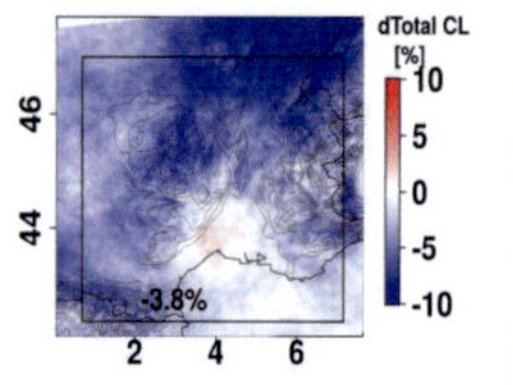	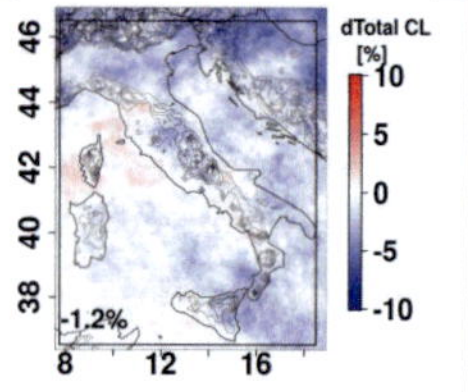

Table 7.9: As Tab. 7.4 for CAPE in ($J\ kg^{-1}$) as obtained by the mean layer method and total cloud cover (CLCT) in [$\%$].

resolution types arise from the different spatial distribution of precipitation and the different simulation boundaries in the 2.8 km runs.

Impact on large-scale pressure and cloud cover

In a like manner to the previous spatial distributions of this section, the geopotential height has been averaged in time for the different climate regions. Likewise Fig. 7.2 and Fig. 6.3, the geopotential height at 500 hPa has been averaged in time and represented in a spatial distribution in order to obtain the Probability Distribution Function (PDF) of this spatial distribution. Figure 7.14 shows the differences (AS-CTRL) of the PDFs between the nudged runs and their respective CTRL over FR. It can be seen that for the 7 km differences larger geopotential height values are more probable in the AS-7 runs as than in CTRL-7. The increase of probability takes place in the range 571 to 572 *gpdm*. This effect has only been observed in the AS-7 simulation over FR. No differences in the Probability Distribution Function (PDF) of the seasonally averaged geopotential height for the 2.8 km simulations are observed. Neither for IT or IP in the AS-7 runs. The reason is that under cooling or drying of the atmospheric column, surface pressure builds up, lifting accordingly the geopotential height at 500 hPa. Hence, given the generalized humidity reduction of the FR region in the AS-7 runs, higher gepotential heights are favoured.

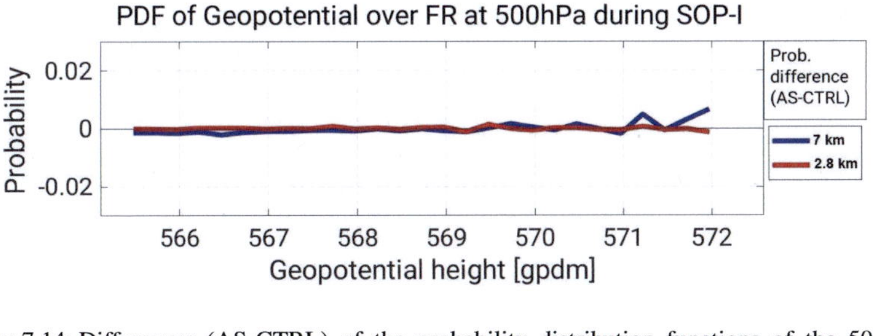

Figure 7.14: Differences (AS-CTRL) of the probability distribution functions of the 500 hPa geopotential height. The PDFs have been obtained from the spatial distribution of the timely averages of FI(500) for the whole period. The blue line accounts for the differences in probability between AS-7 and CTRL-7 and the red for the differences between AS-2.8 and CTRL-2.8

The fact that only over FR, AS-7 shows this impact can be explained by two factors. First, FR has the largest GPS station coverage of the three climate regions. Second, during autumn, low pressure systems typically reach FR advected from the west. These are usually extratropical cyclones, polar cut-off lows or troughs. Thus, when the column atmospheric moisture nudging dries the atmosphere over France, the intensity of these low pressure systems is weakened and pressure at the surface and the geopotential height at 500 hPa are increased. In spite the fact that the differences in probability of larger geopotential heights are small, it should be considered that they account for timely and spatially averages of this quantity. Besides, an in-depth study of a heavy precipitation event of the period has shown that such weakening of the low pressure systems weakens the associated wind circulation, and the moisture transport. Finally, the fact that this impact is not observed in any 2.8 km simulation is due to the smaller simulation boundaries (see figure 5.2).
Provided the shown changes due to the moisture nudging on instability and surface fluxes, seasonally averaged cloud cover experiences a strong impact. A strong reduction of the cloud cover (of ca. -4 % over all climate regions) takes place in the AS-7 and AS-2.8 runs. This occurs mainly over the European continent and some parts of the Mediterranean Sea (northern part of the basin). Table 7.9 shows the spatial distribution of total cloud cover differences between AS and CTRL for both resolutions. This reduction of cloud coverage occurs for low and medium clouds, associated with convective storms.

7.4 Moisture Effects at the Event-scale. The IOP6 and IOP16 Case Studies

Section 7.3 has shown large impacts of the column atmospheric moisture nudging in convective model variables on a seasonal scale. Large impacts induced by the nudging were assessed for atmospheric instability, conditions at the surface, cloud cover and, even though of weaker intensity, for the large-scale pressure distribution. In the following, the assessment of the impact of the moisture nudging on the processes leading to convection is carried out for two particular HPEs of the 1-Sep to 20-Nov period. The simulations of the autumn period using COSMO-CLM on a 7 km and a 2.8 km resolution are employed for event-scale analysis of the the two selected HPEs. For details on the simulations used refer to Tab. 5.2.

The first event is the HyMeX IOP16, which took place over Corsica in the evening of 26-Oct-2012 and early night of the 27-Oct-2012 and the second is HyMeX IOP6, over the Cévennes-Vivarais (CV) region (southern France) on the night of 24-Sep-2012 (see Fig. 3.4). Precipitation totals of 100 *mm* in the course of 12 h were measured by rain gauges over the affected regions for both HPEs. These two case studies fall into the category IWV-increase and precipitation-decrease (IOP16) and the IWV-decrease and precipitation-decrease (IOP6) from the *IWV-precipitation relationship* method presented in Section 7.2 (see Table 7.2). They are selected given their large sensitivity to the humidity changes, in part because they are weakly synoptically forced. Hence the question, *What is the impact on the processes leading to convection, that induced a decrease in precipitation in both cases in spite of reducing IWV in the preconditioning of IOP16 and an increasing it in IOP6?*

Changes in moisture transport leading to orographic precipitation reduction. IOP16

IOP16 took place over Corsica between 26-Oct 10:00h and 27-Oct 16:00h, where precipitation intensities as large as 20 $mm\ h^{-1}$ were measured by rain gauges over the island. Other affected regions were CV, LT and NEI (see Fig. 5.2). Over Corsica, the HPE was orographically lifted and had a weak synoptic forcing. On the 26-Oct, a large-scale low pressure system was slowly moving eastward over the Iberian Peninsula and France (see figure 7.15) with associated southwesterly inflow at the lowest atmospheric heights, bringing warm and moist air from lower latitudes (Martinet et al., 2017). Several potential vorticity anomalies moved through the southwesterly diffluent flow over

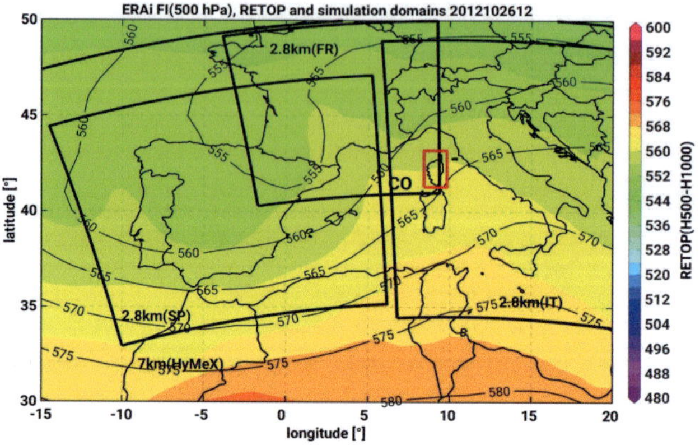

Figure 7.15: ERA-interim reanalysis, relative topography (H500-H1000) in colour shading and geopotential height at 500 hPa (black contours) on the 26-Oct 12:00h. The synoptic situation is dominated by a slow moving upper-level low over the Iberian Peninsula and France inducing a moist southwesterly flow.

the Iberian Peninsula and high values of the wet bulb temperature ($\theta_w < 16°C$) were measured at the French coasts due to the incoming winds (Duffourg et al., 2016). A combination of low-level convergence, orographic lifting and propitious atmospheric instability and moisture conditions, favoured the befallen of convective precipitation. Precipitation over Corsica dissipated during the 27-Oct prior to a change of the displacement direction of the upper-level low toward southern Italy, initiating precipitation across the Italian peninsula.

Figure 7.16 shows the temporal evolution of spatially averaged IWV (dotted lines) and hourly precipitation (straight lines) over CO for the four simulations (colours). The figure shows that two maxima of heavy precipitation take place for the CTRL-7 and AS-7 runs with intensities of 50 $mm\ d^{-1}$ for the first maxima and 24 $mm\ d^{-1}$ for the second. For its part, the CTRL-2.8 runs showed three maxima (42 $mm\ d^{-1}$, 70 $mm\ d^{-1}$ and 24 $mm\ d^{-1}$) which were reduced (especially the second peak) by the moisture nudging (see figure 7.15). Figure 7.16 also demonstrates how AS-7 and AS-2.8 runs represented up to 25 $mm\ d^{-1}$ until 24-Oct 10:00h. This belated precipitation in the AS runs is responsible for their increased IWV amount until 26-Oct 10:00h. This precipitation period corresponded to IOP15. The comparison against the temporal evolution of the spatially averaged precipitation measured by rain gauges over the island showed that the COSMO-CLM represented well the timing of the event and the range of precipitation intensities. The impact of the moisture nudging, however did not bring an appreciable improvement in the representation of precipitation amount.

Regarding IWV temporal evolution, there are three differentiated periods of the impact of the column atmospheric moisture nudging. The first period, between 24-Oct 00:00h and 26-Oct 00:00h is characterized by the aforementioned IWV increase in the AS runs (ca. 10 mm larger). In the second period the differences in IWV between the AS and the CTRL runs, progressively diminish, as IWV increased over the area. As will be shown further in this section, the IWV increase during this period was due to the arrival of moisture originating at the Atlas mountains (north Africa). The last period lasts from 26-Oct 13:00h to 27-Oct 06:00h, when IWV is reduced by about 5 mm in the AS runs as compared to their CTRL counterparts. This last period is that of precipitation.

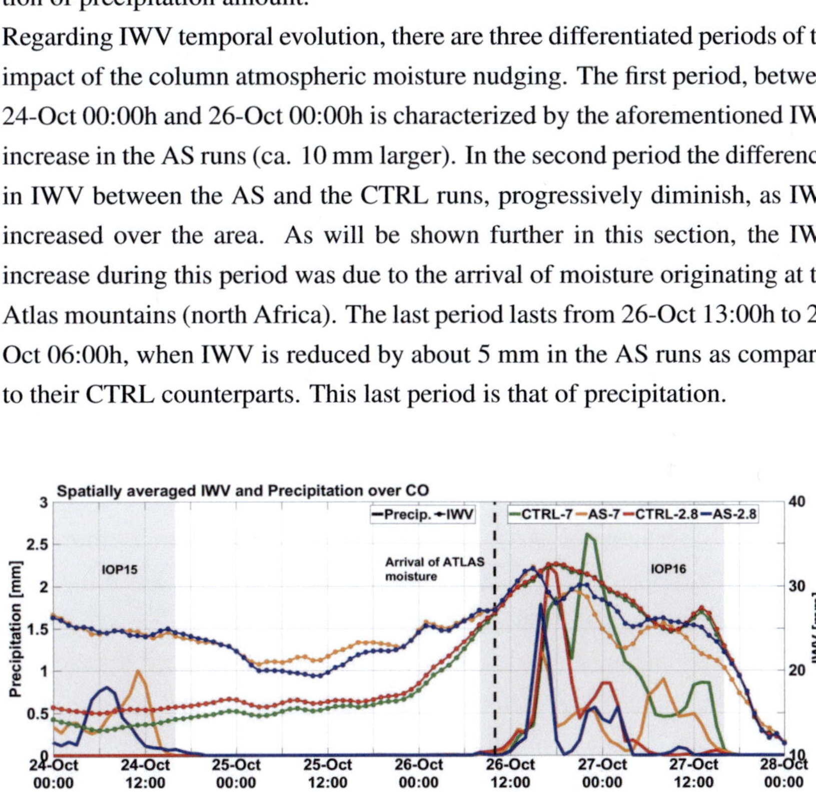

Figure 7.16: Temporal evolution of spatially averaged IWV (dotted lines) and precipitation (straight lines) over CO. All simulations are represented in colours. The time periods of two HPEs, IOP15 and IOP16 are denoted with gray shadings.

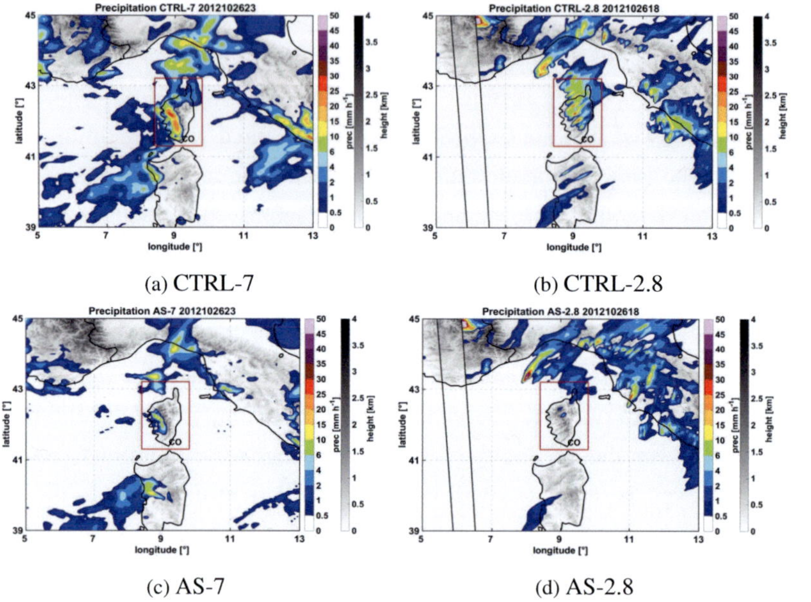

(a) CTRL-7

(b) CTRL-2.8

(c) AS-7

(d) AS-2.8

Figure 7.17: Spatial distribution of hourly precipitation on the 26-Oct 23:00h for CTRL-7 (a) and AS-7 (c) and on the 26-Oct 18:00h for CTRL-2.8 (b) and AS-2.8 (d). The hours of maximum precipitation rate in the CTRL-7 and CTRL-2.8 runs are shown which are delayed by about 5 hours.

The spatial distribution of hourly precipitation at its maximum (see Fig. 7.17) for CTRL-7 (26-Oct 23:00h) and CTRL-2.8 (26-Oct 18:00h) shows the location of precipitation at the windward side of the mountain system. The moist low-level flow impinges over the mountain slopes being forced to reach the LFC evolving to deep convective clouds. The impact of the column moisture nudging is to reduce the maximum precipitation intensity from ca. 30 $mm\ h^{-1}$ (CTRL-7) to 15 $mm\ h^{-1}$ (AS-7) and from 30 $mm\ h^{-1}$ (CTRL-2.8) to 6 $mm\ h^{-1}$ (AS-2.8). Regarding the CTRL-7 to CTRL-2.8 comparison, differences in the precipitation structure exist, specially ahead of the Corsican coast since CTRL-7 represents intense precipitation induced by low-level convergence. The second precipitation maximum (see Fig. 7.16) is brought about (in the CTRL-7 and and AS-7 simulations at 27-Oct 10:00h) a late intensification of the southwesterly flow prior to the veering of the pressure low southward. On the contrary, the second precipitation maximum in the CTRL-2.8 and AS-2.8 runs (27-Oct 00:00h) stems from a transition of the orographic precipitation to the lee side of the mountain (not shown).

Heavy precipitation over CO during IOP16 is supported by the arrival of a high moisture amount originating at the Atlas mountains (north Africa) several days before (from the 21-Oct to 25-Oct). Evaporation of 1.5 $kg\ m^{-2}$ per day over northern Africa took place between 21-Oct and 25-Oct as shown by the daily

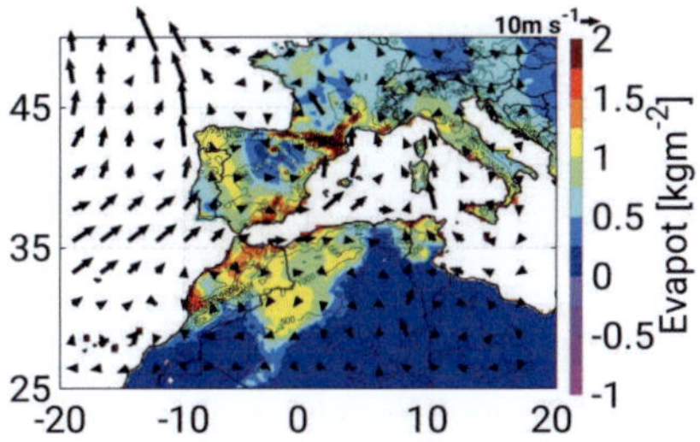

(a) 21-Oct (GLEAM)

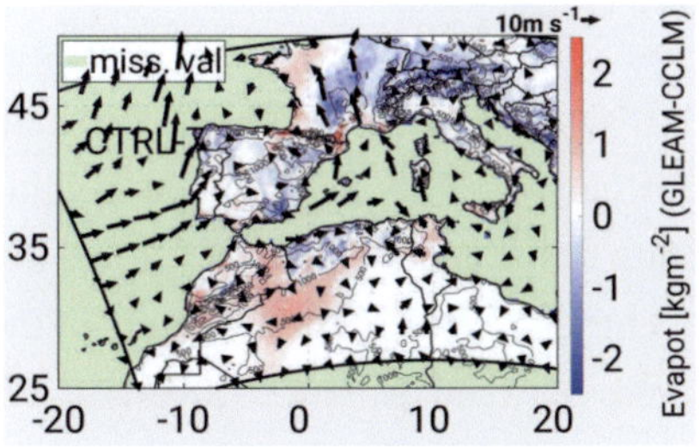

(b) 21-Oct (GLEAM vs CTRL-7)

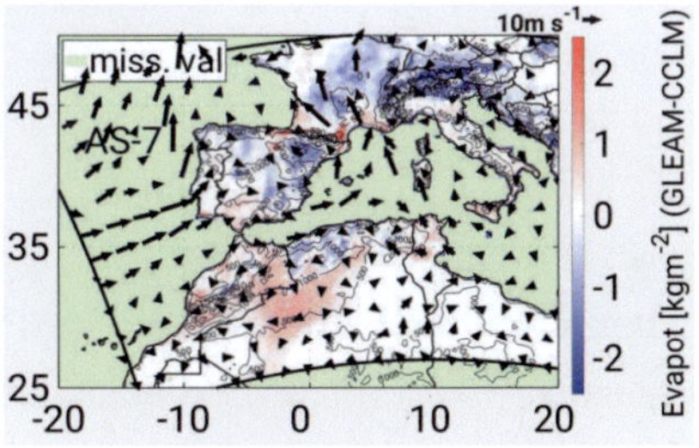

(c) 21-Oct (GLEAM vs AS-7)

Figure 7.18: (Top) daily evapotranspiration over land by the Global Land Evaporation Amsterdam Model (GLEAM) product (see Chap. 5 for a description of the data set) and the differences against CTRL-7 (b) and AS-7 (c) for the 21-Oct-2012 (bottom). The wind fields in a) are obtained from ERA-interim reanalyses and in b) and c) from CTRL-7 and AS-7 respectively. The COSMO-CLM fields have been upscaled to the GLEAM grid of 0.25°x0.25°.

evapotranspiration product of the Global Land Evaporation Amsterdam Model (GLEAM) product. Fig. 7.18 shows this intense evapotranspiration on the 21-Oct. The days until 25-Oct can be found in the Annex. COSMO-CLM is able to represent this large evapotranspiration although with differences down to 1.5 $kg\ m^{-2}$ less in the CTRL-7 and AS-7 runs (see Fig. 7.19b and Fig. 7.19c). COSMO-CLM is able to represent this intense moisture evaporation in all simulations. The moisture originating at the Atlas mountains reaches the Mediterranean Sea on the 25-Oct-2012. This is shown by the Moderate Resolution Imaging Spectroradiometer (MODIS) product and CTRL-7 and AS-7 also show this moist air mass (see Figure 7.19). This day is shown, as the day where the Atlas moisture was advected over the Mediterranean Sea toward Corsica and the Liguria-Tuscany region. The results for CTRL-2.8 and AS-2.8 are not included given their narrower simulation domains.

Once the very humid Atlas air mass reaches Corsica, a maximum of spatially averaged IWV over the island takes place on the 26-Oct 14:00h in all simulations. Later in the day, the AS runs started to diverge from their CTRL counterparts reducing IWV by about 5mm for the following 12h.

At 26-Oct 13:00h the spatial averages of IWV for AS-7 and AS-2.8 start to diverge from those of CTRL-7 and CTRL-2.8, reaching clear differences at 26-Oct 18:00h of ca. 5mm less. This brings an immediate impact on latent instability that, together with the reduced moisture flux, at different heights over the island impedes stronger convection development. Figure 7.20 shows in box and whiskers plots the median, percentiles 25 and 75 and extremes of the simulated values for several model variables between 26-Oct 09:00h and 27-Oct 16:00h (i.e. during precipitation). Extreme CAPE is reduced from ca. 675 $J\ kg^{-1}$ to 310 $J\ kg^{-1}$ in the 7 km runs and from ca. 800 $J\ kg^{-1}$ to 710 $J\ kg^{-1}$ in the 2.8 km runs. In a like manner, specific humidity is reduced between 700 hPa and 950 hPa steadily (not shown) with large variations of ca. 0.5 $g\ kg^{-1}$ at 700 hPa (both resolutions) and at 950 hPa (most notably in the 7 km runs). Convection is, thus, hindered given the buoyancy reduction of the atmospheric column, in addition to a drier lower mid-troposphere. This, favoured dry air

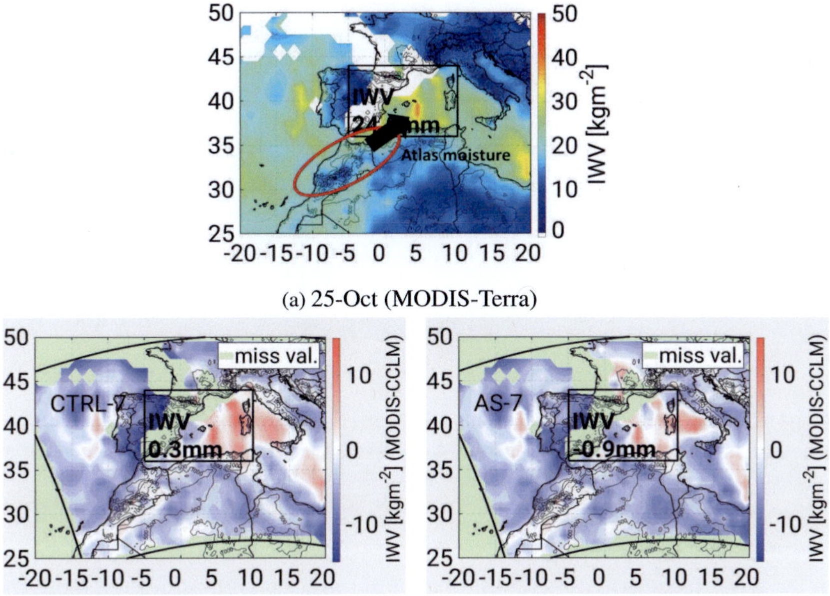

(a) 25-Oct (MODIS-Terra)

(b) 25-Oct (MODIS vs CTRL-7)

(c) 25-Oct (MODIS vs AS-7)

Figure 7.19: Daily averaged IWV spatial distribution on the 25-Oct-2012 as measured by the Moderate Resolution Imaging Spectroradiometer (MODIS) instrument aboard the Terra satellite (top) and differences against CTRL-7 (bottom left) and AS-7 (bottom right). The COSMO simulations have been upscaled to the MODIS 1°x1° grid to allow for comparison

intrusions in the convective plume and the detrainment of water vapour through the cloud boundaries to the environment. This is the case for both resolutions. Figure 7.20 also gives a quantification of the moisture flux values and wind speeds simulated during the event. Moisture flux is reduced in agreement with the specific humidity and IWV reductions addressed earlier in this section. Values of median moisture flux, change from ca. 72 $g\ m\ kg^{-1}\ s^{-1}$ (CTRL-7) to 63 $g\ m\ kg^{-1}\ s^{-1}$ (AS-7) and from 72 $g\ m\ kg^{-1}\ s^{-1}$ (CTRL-2.8) to 68 $g\ m\ kg^{-1}\ s^{-1}$ (AS-2.8). The changes in moisture flux between CTRL-7 and AS-7 are smaller than their 2.8 km counterparts, this is due to the changes

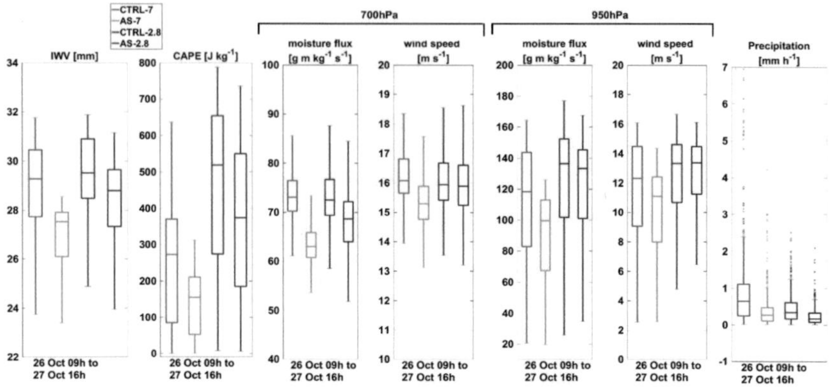

Figure 7.20: Box and whiskers plots showing the median (mid horizontal line), the percentiles 25 and 75 (the box edges) and the extreme values (whiskers). All box and whiskers are obtained from the distribution of values for the shown quantities between the 26-Oct 09:00h and 27-Oct 16:00h over the study region CO.

in the wind speed and direction in the 7 km runs. The reason for the modified wind field in the AS-7 run as compared to CTRL-7 is the impact of the moisture nudging on the low pressure system located over France during the event (see Fig. 7.21). The reduction of wind speeds in the AS-7 run took place at all represented pressure levels bringing changes of 1 $m\ s^{-1}$ in the median at 700 hPa and 950 hPa.

During IOP16 (see figure 7.15) a low pressure system over southern France and the Iberian Peninsula induces a cyclonic southwesterly flow from 700 hPa down to 950 hPa. Figure 7.21 shows the differences of Pressure at the Mean Sea Level (PMSL) between the AS-7 and CTRL-7 runs so that red accounts for higher PMSL in the AS-7 as compared to CTRL-7. Additionally, Fig. 7.21 shows horizontal winds at 950 hPa by AS-7 (yellow arrows) and CTRL-7 (green arrows). A clear increase of PMSL at the center of the system occurs as a result of the moisture nudging (of ca. 10 hPa) as well as a reduction of the width. This causes a weakening of the low and of the associated wind circulation, down to the lowest tropospheric levels. The PMSL anomaly is already

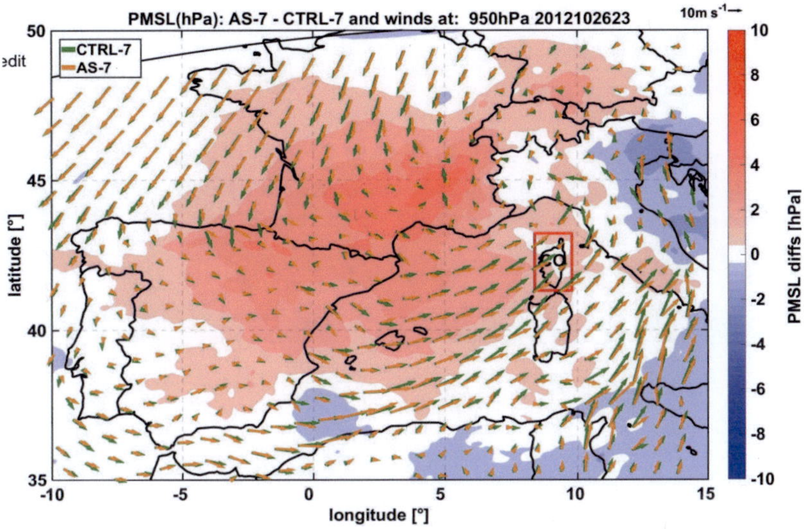

Figure 7.21: Spatial distribution of the differences in Pressure at the Mean Sea Level (PMSL) between AS-7 and CTRL-7 on the 26-Oct 23:00h. Horizontal winds at 950 hPa are represented by green (CTRL-7) and yellow (AS-7) arrows. The column atmospheric moisture nudging in AS-7 weakens the upper-level trough over FR and the associated wind circulation.

visible at 25-Oct 15:00h and persists as the trough reaches southern France. The increase of PMSL was caused by a strong humidity decrease induced by the moisture nudging over FR and IP with reductions down to 6mm in spatially averaged IWV lasting between 25-Oct and 26-Oct (not shown).

This effect on wind circulation of the column atmospheric moisture nudging was not observed in the AS-2.8 simulations given the narrower simulation domains.

Due to the aforementioned impacts on instability (see fig 7.20), detrainment (Fig. 7.20), moisture feeding (Fig. 7.19a) and dynamic triggering (only in the AS-7 runs) the represented convection was weaker on both AS-7 and AS-2.8. Figure 7.22 represents the vertical winds at 500 hPa in a red blue scale over the study region, for all simulations on the days of most intense convection

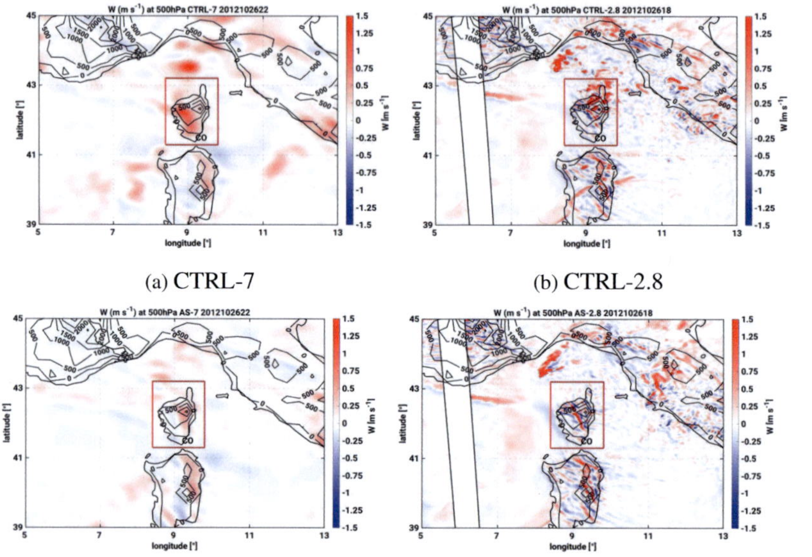

(a) CTRL-7 (b) CTRL-2.8

(c) AS-7 (d) AS-2.8

Figure 7.22: Spatial distribution of vertical wind speed on the 26-Oct 22:00h (CTRL-7 and AS-7) and on the 26-Oct 18:00h (CTRL-2.8 and AS-2.8). Red colours stand for more intense updrafts in the AS as compared to CTRL. Less convective updrafts are simulated in AS-7 and AS-2.8 as a result of the humidity changes in the moisture source and of dynamic lifting (7 km).

in the CTRL-7 and CTRL-2.8 runs (26-Oct 22:00 and 26-Oct 18:00h respectively). A remarkable decrease in the intensity of the convective updrafts can be seen at this height for both resolutions. AS-7 shows shallower and less intense updrafts (about 0.75 $m\ s^{-1}$ less) than its CTRL counterpart. Similarly, AS-2.8 represented, less updrafts, of weaker intensity and only over the mountain slopes of Corsica. Whereas CTRL-2.8 also showed convection over the sea northwest to the island (see figure 7.22).

Reduced PBL moistening hindering instability build-up. IOP6

This event took place during the night of 24-Sep-2012 and left precipitation amounts of 100 mm in 12 h over southern France and North-eastern Italy. This HPE is selected for in-depth study since, as opposite to IOP16, the effect of the column atmospheric moisture nudging on precipitation showed the same sign of change as IWV. For IOP6, the nudging of column moisture induced a reduction of humidity prior to the initiation of the event that resulted in a precipitation decrease in the AS-7 and AS-2.8 runs.

The synoptic situation during IOP6 was dominated by the arrival of a westerly upper-level trough associated with low surface pressure northwest of Ireland on the 23-Sep and 24-Sep (Caldas-Álvarez et al., 2017). The trough induced a southwesterly wind circulation bringing in moist and warm air from the Mediterranean to southern France (see figure 7.23), and the Gulf of Genoa. Triggering took place at the arrival of an approaching frontal discontinuity with enhanced baroclinicity (Dafis et al 2017). At about 24-Sep 00:00h, given the

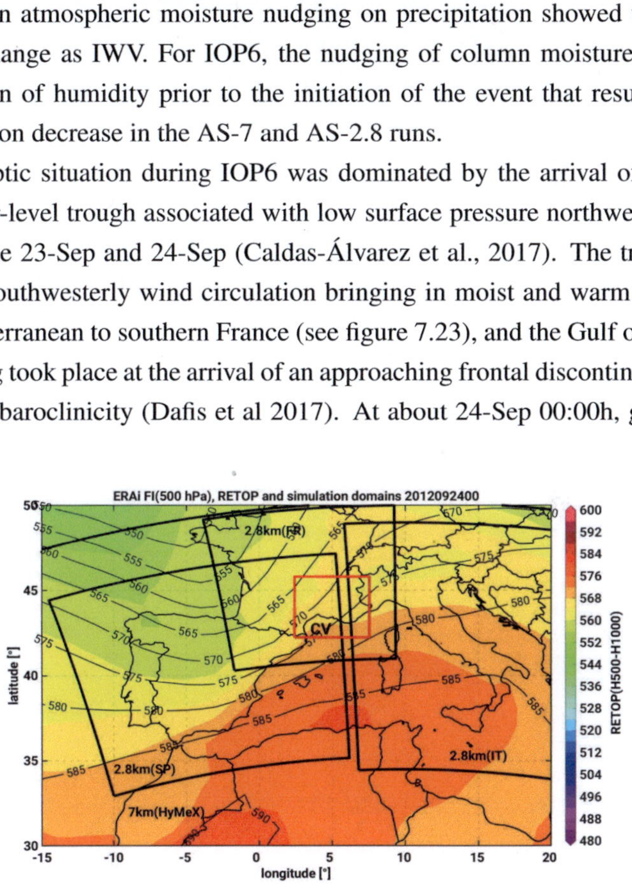

Figure 7.23: ERA-interim reanalysis, relative topography (H500-H1000) in colour shading and geopotential height at 500 hPa (black contours) on the 24-Sep 00:00h. Upper-level trough-ridge pattern bringing an intense moistening of the low levels to the CV area.

large shear associated with the trough, convection organized into the form of a large MCS over southern France reaching the Alpine region at about 24-Sep 09:00h. A clear linear convective signature could be observed both in the radar imagery and the rain gauges (Hally et al., 2014) and intense lightning swept the area between 24-Sep 00:00h and 24-Sep 09:00h with about 5000 cloud to ground strokes every 2h hours (Defer et al., 2014; Dafis et al., 2017). This HPE is particularly suitable for the assessment of the nudging impact, given the combination of the local factors building instability and providing moisture with a large-scale disturbance triggering convection.

Figure 7.24 shows the temporal evolution of the spatially averaged IWV (dotted lines) and precipitation (straight lines) over the CV area between the 22-Sep 12:00h and the 25-Sep 00:00h. It can be seen how precipitation is initiated for all simulations at 23-Sep 21:00h and lasts for about 19h. Maximum precipitation shown by CTRL-7 after averaging in space reaches 30 $mm\ d^{-1}$ on the 24-Sep 09:00h; 18 $mm\ d^{-1}$ in the AS-7 on the 24-Sep 07:00h with; 30 $mm\ d^{-1}$ in the CTRL-2.8 on the 24-Sep 08:00h and 24 $mm\ d^{-1}$ in the AS-2.8 on the 24-Sep 08:00h. Likewise for IOP16, the comparison against the rain gauges

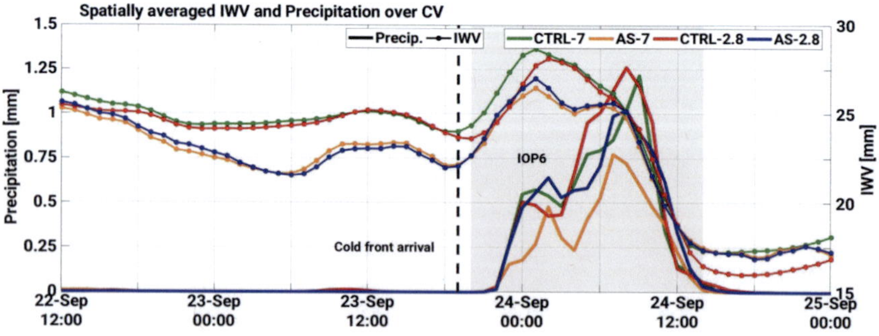

Figure 7.24: Temporal evolution of spatially averaged IWV (dotted line) and precipitation (straight line) over CV. All simulations are represented in colours. The period of precipitation during IOP6 is highlighted. Both AS-7 and AS-2.8 show less IWV in the preconditioning of the event inducing less precipitation during the event.

showed that COSMO-CLM reproduced well the features of the event but that the column moisture nudging could not bring an improvement.

Regarding IWV, differences of ca. 3mm in the spatial averages IWV exist between the AS and CTRL runs of both resolutions between 22-Sep 18:00h and 23-Sep 20:00h and also between 24-Sep 02:00h and 24-Sep 06:00h. A high IWV amount was present over the area already from the 21-Sep 00:00h on, due to a persistent moisture flux with the southwesterly wind. Regarding the comparison of the IWV representation of the 7 km grid and the 2.8 km, very similar IWV amounts were simulated.

Figure 7.25 shows the spatial distribution of precipitation on the 24-Sep 02:00h for all simulations. It can be observed how COSMO reproduces well the convective line signature assessed in previous studies in all of them. In spite of

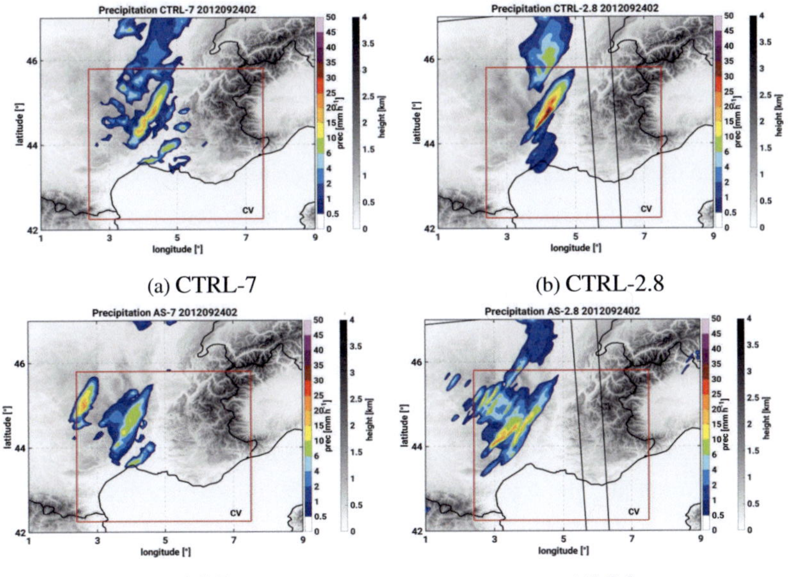

(a) CTRL-7 (b) CTRL-2.8

(c) AS-7 (d) AS-2.8

Figure 7.25: Spatial distribution of hourly precipitation on the 24-Sep 02:00h (left, CTRL-7 and AS-7 and right, CTRL-2.8 and AS-2.8).

the similar line shape of the precipitation field, relevant differences exist in the structure of the precipitation fields and the maximum precipitation rate among simulations. At the shown time, CTRL-7 and AS-7 represent a maximum precipitation rate of about 25 $mm\ h^{-1}$ with maximum precipitation taking place along the valley in the former and over the Massif Central in the latter. Similarly, CTRL-2.8 shows the convective line over the valley with maximum precipitation rate of ca. 35 $mm\ h^{-1}$ and AS-2.8 shows precipitation spread toward the Massif Central precipitation with maximum intensities of 25 $mm\ h^{-1}$.
On the 21-Sep, persistent moisture flux, reached southern France and bundled along the Avignon valley and the Gulf of Lion. As shown by figure 7.24, spatially averaged IWV for the CTRL and AS simulations starts to diverge on the 22-Sep 18:00h with a reduced IWV amount in the AS runs that continues until they are largest on the 23-Sep 06:00h. The box and whiskers plots presented in figure 7.26 quantify the median, percentile 25 and 75 values and the extremes of several convective-related quantities between the 23-Sep 06:00h (the moment where IWV differences are largest during the preconditioning) and the 24-Sep 10:00h (the end of precipitation over CV).
Figure 7.26 demonstrates the decrease in the median, the quartiles, and extreme IWV in AS-7 and AS-2.8 as compared to their CTRL counterparts. As a result, CAPE is lowered over the area with reductions of ca. 50 $J\ kg^{-1}$ in the median and percentile-75 values. Furthermore, CIN is increased by ca. 50 $J\ kg^{-1}$ in the AS-7 and AS-2.8 in the median, reaching -750 $J\ kg^{-1}$ for the largest CIN in both AS runs. KO-index median values are lower for the CTRL runs (ca. -4 K) than for the AS runs (ca. -3 K) indicating more potential instability. Provided the decreased conditional instability in the AS-7 and AS-2.8 simulations, the larger inhibition and the weaker potential instability, extreme precipitation is reduced in its median, percentile-75 and extremes. Regarding the comparison between 7 km and 2.8 km, it can be seen how for CAPE, CIN and KO-index the range of represented values is very similar between the AS-7 and the AS-2.8 runs.

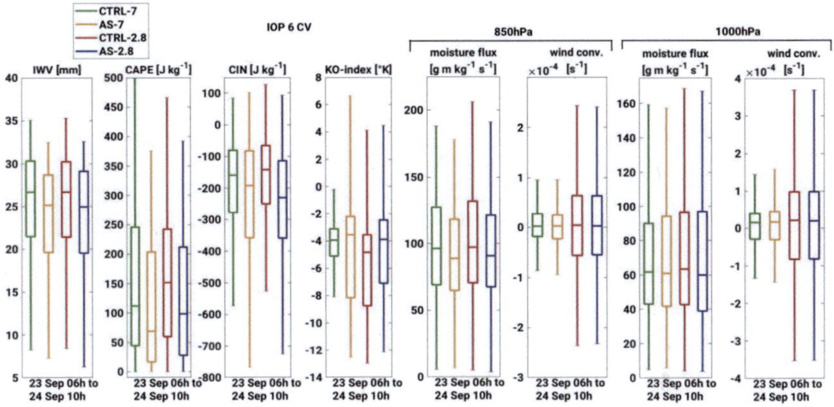

Figure 7.26: Box and whiskers plots showing the median, percentile 25 and 75 and extremes of IWV, CAPE, CIN and KO-index, as 2D atmospheric fields and moisture flux and wind convergence at 850 hPa and 1000 hPa during the period 23-Sep 06:00h to 24-Sep 10:00h

Figure 7.26 shows, as well, the median, quartiles and extreme values for moisture flux and wind convergence at 850 hPa and 1000 hPa. In agreement with the reduction of IWV from 500 hPa (not shown) down to 1000 hPa, median specific humidity is reduced ca 0.5 $g\ kg^{-1}$ at 850 hPa in the AS runs of both resolutions and ca. 0.25 $g\ kg^{-1}$ at 1000 hPa. Finally, winds were only marginally affected in their speed by the moisture nudging, and indeed wind convergence at 1000 hPa only showed minor differences. Convergence over the area is caused by the impinging of the westerly flow, behind the cold front on the Alpine mountains. More intense wind convergence occurs in the 2.8 km runs as compared to the 7 km due to the intensification of the wind speeds in convection permitting simulations as assessed by past studies (Barthlott and Hoose, 2015).

Given these strong impacts on instability and moisture flux over CV during IOP6, the intensity, the number of cells triggered and the depth of convection was hampered by the column atmospheric moisture nudging. Figure 7.27 shows the vertical winds at 500 hPa for all simulations on the night of 24-Sep at 05:00h. At this time, mature convective cells are present over CV for

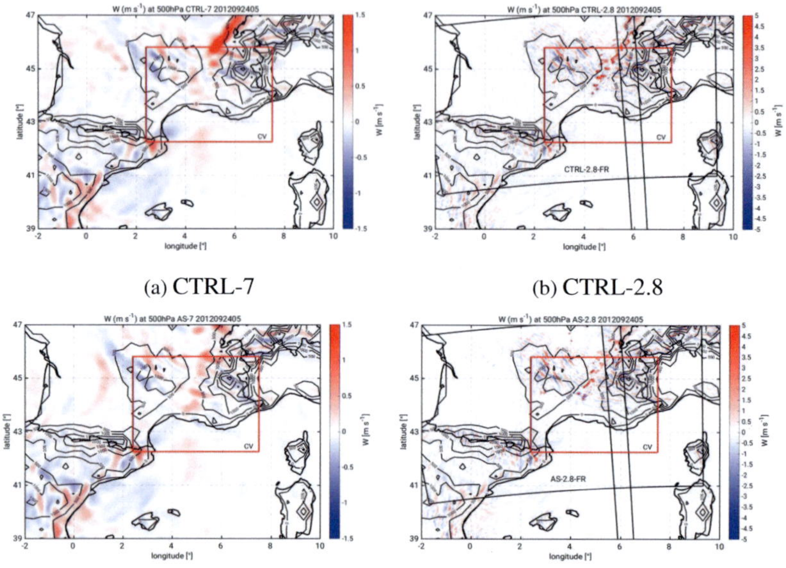

(a) CTRL-7 (b) CTRL-2.8

(c) AS-7 (d) AS-2.8

Figure 7.27: Spatial distribution of vertical wind speed at 500 hPa on the 24-Sep 05:00h. All simulations present a convective line structure along the Avignon valley whose updrafts are represented weaker, shallower and in less number in the AS runs.

all simulations and differences in upward vertical winds are of the order of $1\ m\ s^{-1}$ between CTRL-7 and AS-7 and of $0.5\ m\ s^{-1}$ between CTRL-2.8 and AS-2.8. Also convective updrafts are less spread in the AS-7 runs. For its part, AS-2.8 showed less convective updrafts than CTRL-2.8 at 500 hPa. It is also noteworthy the intensified wind updrafts shown by the 2.8 km simulations in comparison to the 7 km grid, maximum wind speeds are three times larger in the finer resolution grid.

In order to assess the accuracy COSMO-CLM in redistributing the atmospheric moisture after nudging, a comparison of the simulated vertical profiles of all runs with an operational sounding located at Nimes (south of the investigation area; see figure 7.23) is presented. For this comparison, all simulation values

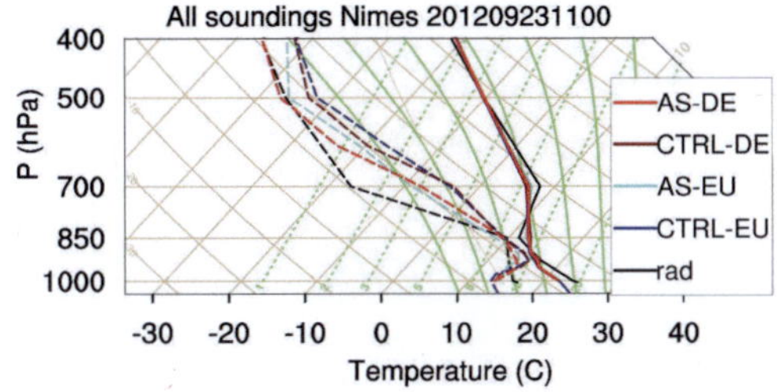

Figure 7.28: Skew-T log-P diagram at Nimes (Southern France; see figure 7.23) on the 23-Sep 11:00h. The operational sounding is shown in black and the profiles obtained by the different simulations are given in colours. The moisture nudging shows a good tendency for correction drying the layer between 400 hPa and 850 hPa.

have been interpolated to the same pressure levels and the grid-points closest to Nimes have been selected. The profiles on the 23-Sep 11:00h, during the preconditioning of the event, show the consistent humidity reduction for the whole atmospheric column in AS-7 and AS-2.8, from 400 hPa down to 1000 hPa. However at the lowest level (1000 hPa) differences in dew point temperature and temperature of the dry bulb are practically the same for all resolutions. Given the larger lapse rate in the CTRL runs, more CAPE is present as compared to the AS. Convection is hampered due to having less energy available for convection but also through enhancement of detrainment between 850 hPa and 400 hPa, where the moisture content is lower in the AS runs.

The column atmospheric moisture nudging induces a good tendency for humidity corrections between 500 hPa and 850 hPa. However, relevant differences between the observational sounding and the AS profiles persist at that layer. Differences in mixing ratio of -1.5 $g\ kg^{-1}$ can be found at the 700 hPa level between the AS runs and the sounding. This reduced humidity at the middle lower troposphere indicates that even if the moisture nudging corrected humidity with the right tendency (i.e. drying of the atmospheric preconvective environment) too large differences exist in the representation of the atmospheric profile that hamper the representation of convection over the area. This might explain the worse representation of precipitation amounts in the AS runs as compared to CTRL.

Conclusions

Q.1 *How well does the COSMO model represent extreme precipitation during the HyMeX-SOP1? Did the nudging of column atmospheric moisture bring an improvement? How did this impact affect the several features of precipitation on the 7 km and 2.8 km horizontal resolutions? (spatial distribution, temporal evolution, representation of extremes, diurnal cycle, etc.))*

COSMO in a near-operational configuration represents well the main regions prone to heavy precipitation in the western Mediterranean. The spatial distribution of the frequency of precipitation above the percentile-90 (see figure 7.6) highlights the Balkan and Italian occidental coasts, Corsica, the Alps, the Pyrenees, the Atlantic Iberian coast, the Atlantic Ocean and Aquitaine and Poitou-Charentes as prone to heavy precipitation (see Fig. 3.4 in Chap. 3.1). The column moisture nudging improved the representation of these extremes over the Mediterranean and the Atlantic (enhancing ca. 6 % the frequency of heavy precipitation) and over the Iberian Peninsula and France (through a decrease of ca. 8 % of the frequency). Still, the column moisture observations were unable to sufficiently increase the number of extreme precipitation days over the Balkans.

For its part, the mean seasonal precipitation values showed improvement in the representation of the precipitation structure over all regions, after nudging column atmospheric moisture. However, the amount, as quantified by the Structure Location Amount (SAL) method, only improved over FR for both resolutions, over IT and IP, the reduction of precipitation brought worse scores. Regarding the representation of the temporal evolution of heavy precipitation, COSMO represented well the timing of the events, however their amount showed relevant differences against observations (see Figs. 7.4 and 7.5).

The column atmospheric moisture nudging induced strong changes, bringing equally increases and decreases of precipitation during the events which rise up to >100 % noticeable at climate and local spatial scales for some HPEs of the period. The moisture nudging improved the representation of the heavy precipitation amount for some HPEs (especially at CA and VA) which showed not to be systematic. Hence the relevance of understanding the role played by the characteristic features of the affected regions (orography, soil type and vegetation, mesoscale circulations, etc.) and of the atmospheric conditions particular of each event.

COSMO showed deficiencies in representing correctly the precipitation distribution function of the SOP1 period over the climate regions when using a 7 km grid horizontal resolution (see figure 7.7). Using a 2.8 km grid increased the probability of precipitation rates above 20 $mm\ d^{-1}$ bringing a better agreement with observations for all climate regions and target areas. This enhancement of the probability of extreme precipitation rates in the 2.8 km simulations, stems from the fact that this grid resolution can represent very localized and large intensities even for events which in mean showed only moderate precipitation totals. Conversely, the use of a 7 km grid and the parameterization of deep convection implies that the largest mean precipitations scale with the largest maximum precipitation rates (see figure 7.10). The improvement was reinforced over IT and IP by the nudging of column moisture observations which topped the increase in frequency for extreme precipitation, bringing a more realistic representation of the precipitation spread. At a local scale, the impact of the column atmospheric moisture nudging was, generally, to lower the probability of precipitation rates larger than 10 $mm\ h^{-1}$ (CTRL-7) and 20 $mm\ h^{-1}$ (CTRL-2.8) but some areas experienced a much more variable impact. Such is the case of the Italian target areas NEI, CI and LT, which showed only minor changes and of CO which event underwent an increase in probability of precipitation larger than 10 $mm\ h^{-1}$.

Another aspect where combining a convection-permitting grid with the nudging of atmospheric moisture was beneficial was the representation of the precipitation diurnal cycle over the climate regions (see Figure 7.7). The nudging of column atmospheric moisture strengthened the afternoon maxima and attenuated the early morning precipitation of all simulations (all climate regions and both horizontal resolutions). For its part, using a 2.8 km grid, rose precipitation amount, aiding the simulations to represent a diurnal cycle closer to remote sensing observations (CMORPH).

The overall impact of the moisture nudging was to correct the IWV wet bias, ultimately reducing precipitation over land (between -65 % and -9 % less precipitation depending on the grid size and the climate regions) for both resolution types. A weaker decrease took place over IT (especially in the 2.8 km simulations) due to the influence of the southwesterly flow which advects over regions without GPS stations. Over the Italian Sea, a much more patchy structure was induced by the column moisture nudging with increases and decreases

Q.2 *What is the role of the IWV variations preceding heavy precipitation? Are their extent and duration relevant for controlling the intensity of heavy precipitation? Is there a cause-effect relation?*

Supplementary to past studies findings, reporting intense moistening preceding heavy precipitation (Chazette et al., 2015b; Schiro et al., 2016; Khodayar et al., 2018), for the HyMeX-SOP1 simulations, it was found that ca. 90% of HPEs were preceded by an IWV monotonic increase (see table 7.1). In the remaining cases, monotonic IWV decreases occurred at an already very moist area.

The preconditioning of HPEs through moistening of the affected areas was characterized to occur in time intervals of ca. 9h in mean bringing changes of ca. $+4$ *mm* of mean IWV. The nudging reduced the median duration of preconditioning as well as the extent of the IWV monotonic variations to ca. 5 h and $+2$ *mm* respectively (see table 7.1). This was so, given the tendency of AS-7 and AS-2.8 to simulate larger IWV variability and therefore shorter periods of monotonic IWV variations. Moreover, the CTRL simulations showed that the events of larger spatially averaged precipitation ($> 1.5\ mm\ h^{-1}$) showed very large IWV monotonic increases ($> 10\ mm$) and large preconditioning periods ($\approx 30\ h$). This did not occur in the AS runs, again, given the larger IWV variability.

When assessing the extent of the changes in precipitation amount in relation to the extent of the changes in IWV during preconditioning caused by the column atmospheric moisture nudging, it was found that ca. 60 % of the events showed an IWV and precipitation change of the same sign. In other words, that when the nudging induced a decrease/increase in IWV during preconditioning the mean precipitation of the event also showed an decrease/increase (see tables 7.2 and 7.3). This holds for both resolution types. However, the 7 km and 2.8 km runs showed a different response to the column moisture nudging, in that the AS-7 runs showed a favoured case of reduced IWV in the preconditioning and reduced mean precipitation of a certain event (in 41% of the cases). This was not the situation for the 2.8 km runs. Indeed, when categorized by their degree of synoptic forcing, the weakly synoptically forced cases in the AS-7 even showed larger probability (43 %) of the less-IWV and less-precipitation situation. This is due to the generalized humidity reduction, strongest at the lowest troposphere, and the use of the Tiedtke parameterization scheme. This scheme sets a closure assumption through which convection will be triggered only when moisture convergence tends to increase the subcloud moisture content (Doms et al., 2011). Whence, the change in precipitating convection.

Finally, the analysis of the seasonal statistics obtained with the *IWV-precipitation relationship* method and of two selected HPEs (IOP16 and IOP6), revealed that humidity variations, once precipitation has started, are decisive to set the mean precipitation of the event. The analysis on IOP16 and IOP6 revealed how the nudging of column atmospheric moisture, had a decisive effect on precipitation in only 1h to 2h after the change in IWV. This indicates that moistening preceding heavy precipitation has a control on convection initiation and the first hours of the evolution of the event but the changes in water vapour occurring once precipitation has started control the amount of precipitation in the remainder of the event.

Q.3 *Which convective processes (and how much) are affected by the moisture nudging on seasonal and event scales? Are any of these, especially relevant to explain the differences in precipitation? If so, is there any tendency for a particular region or period?*

Two ingredients of convection, instability and moisture supply were, in all cases, impacted by the changes induced by the column moisture nudging. Latent and potential instability values, quantified by CAPE, CIN and KO-index, showed generally less instability over the affected areas when the nudging reduced IWV (see table 7.9). This was demonstrated for the seasonal means as well as for the two selected presented HPEs (see figures 7.20 and 7.26). For its part, the moisture flux reduction hampered deep convection through the processes of detrainment and dry air intrusions. An example of which is the impact of the column moisture nudging in the moisture amount during IOP16 over CO. During this event a very moist air mass reaches Corsica impinging over the windward slopes. The intense evapotranspiration and the advected moist air mass are visible in the GLEAM and MODIS products The column atmospheric moisture nudging reduces humidity in the last 12h of the event, reducing the buoyancy of the air masses and favouring mixing.
The nudging showed also a strong impact on the moist thermodynamic conditions at the model surface. The impact on 2m-temperature, $2m - q_v$ (see Fig. 7.6), surface latent and sensible heat fluxes showed a large spatial variability with increases and decreases. The process chain explaining this heterogeneous impact depends largely on the changes of outbound LH flux, hence on changes in seasonal precipitation. This process chain is explained as follows: given, the overall humidity reduction (also at the surface-atmosphere boundary), water parts and very high mountains (>3000 m) showed intensified LH flux emission. Over lower land points, the regions showing an enhancement of the LH surface outbound flux, were those where the column moisture nudging also induced relevant precipitation increases. In turn, COSMO tried to keep constant the net radiation at the surface, therefore increasing the outbound surface SH

flux, there where outbound LH flux was reduced (low terrain, land points and regions with a dense GPS coverage) in the AS runs. Finally, 2m-temperature was increased at those same regions where outbound SH flux was enhanced.

In the seasonal scale, a positive feedback, between the changes induced by the nudging in humidity (IWV and specific humidity, throughout the complete column), and precipitation is observed. In the process chain less instability is simulated by the AS runs which in turn hampers convection as demonstrated by the decreases in total cloud cover.

The large scale pressure and geopotential height distributions were also affected, although to a weaker extent and only for the 7 km runs. The seasonal means for the geopotential height at 500 hPa (see figure 7.14) as well as the Pressure at the Mean Sea Level (PMSL) for IOP16 (see figure 7.21) showed that the column atmospheric moisture nudging brings an increased probability of larger surface pressure and geopotential height values at 500 hPa. For the autumn period in the WMed a common synoptic situation is that of low pressure systems advecting towards the east, passing over the Iberian Peninsula and France. Under the case of an humidity reduction the low pressure systems undergo a weakening. This, as shown for IOP16 (see Fig. 7.22), affects the associated wind circulation also at the lowest levels, bringing an impact on triggering through dynamic lifting as well as a difference in the redistribution of moisture.

8 Effect of Combined Nudging of Column Atmospheric Moisture and Radiosondes

The analysis on the impact of the column moisture nudging in the simulation of humidity during the HYdrological cycle in the Mediterranean eXperiment (HyMeX)-Special Observation Period 1 (SOP1), see Chap. 6, revealed biases in the vertical distribution of moisture in spite of bringing an improved Integrated Water vapour (IWV) representation. Root Mean Square Error (RMSE) values of specific humidity up to 1.6 $g\ kg^{-1}$ existed at the PBL in the seasonal means of the SOP1 for selected stations and up to 1.5 $g\ kg^{-1}$ at 700 hPa.
Past sensitivity studies have shown the stringent impact of modifying the mixing ratio at chosen pressure levels (Keil et al., 2008; Li and Shen, 2010; Sherwood et al., 2010; Lee et al., 2018). Section 2.3 summarizes state of the art findings on modelling experiments inducting humidity variations of the order of 1$g\ kg^{-1}$ to 2 $g\ kg^{-1}$ at chosen model heights. For instance in Lee et al. (2018) a reduction of 2 $g\ kg^{-1}$ in the marine atmospheric boundary layer upstream of convection induced a decrease of 10 % in total precipitation at the coastal region.
The relevance of a realistic vertical profile representation is addressed by reproducing the Intensive Observation Period (IOP) 6 (see Section 7.4) using very-high resolution simulations (down to a 500 m horizontal resolution grid, reaching micro-α scales) in a Numerical Weather Prediction (NWP) set-up combined with nudging of column atmospheric moisture and operational radiosonde. The rationale behind this modelling experiment is being able to

better represent the fine-scale moisture variations and physical processes, supported by the nudging of atmospheric profiles. The following scientific questions are set out for this chapter.

Q.1 *Can added value be gained in the precipitation and humidity representation of HyMeX-IOP6 by nudging highly sampled humidity observations (GPS-ZTD), jointly with realistic atmospheric profiles and refining the model grid (7 km, 2.8 km and 500 m)?*

Q.2 *What is the impact for the processes leading to convection of nudging column atmospheric moisture observations combined with atmospheric soundings in very high-resolution simulations of IOP6?*

8.1 Improved Representation of Fine-scale Humidity and Precipitation

As presented in section 7.4, HyMeX IOP6 (24-Sep-2012) was characterized by a moist southwesterly to south-easterly flow, bringing large amounts of water vapour inland, through the Avignon and Po valleys and toward the Gulf of Genoa. At the arrival of the upper-level disturbance and the mesoscale front (ca. 24-Sep 02:00h), convection triggered across southern France developing into MCSs moving towards north Italy, regions which exhibited large instability values and very moist low-level air masses. In addition, the mountain ridges of the Alps and Massif Central succeeded in triggering new cells at the foothills. In order to verify the representation of IWV by the different COSMO simulations, Figure 8.1 represents the differences in IWV between the observations (GPS-IWV) and COSMO for the CTRL-7 (a), CTRL-2.8 (c) and CTRL-500 (e) and the AS-GPS-RAD-7 (b), AS-GPS-RAD-2.8 (d) and AS-GPS-RAD-500 (f). The remaining simulations for this experiment (see Tab. 5.3) can be found in the annex in Fig. A.5. The differences shown, stand for the period 23-Sep 12:00h to 23-Sep 19:00h, when the nudging of observations (column atmospheric moisture and radiosondes) brought the large humidity decrease.

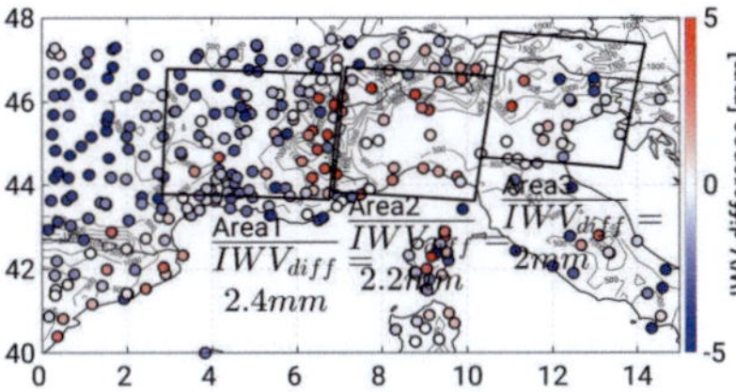

(a) CTRL-7

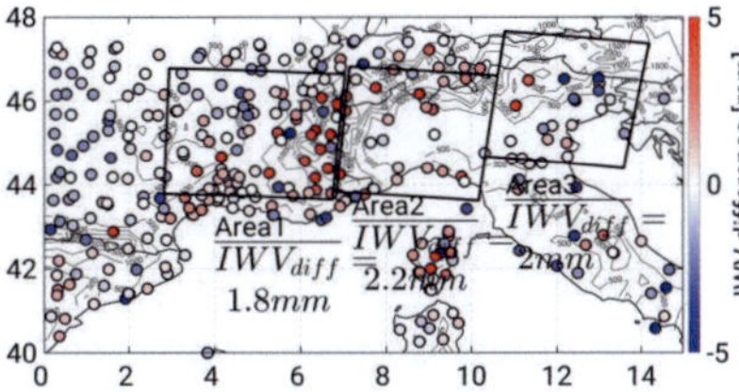

(b) AS-GPS-RAD-7

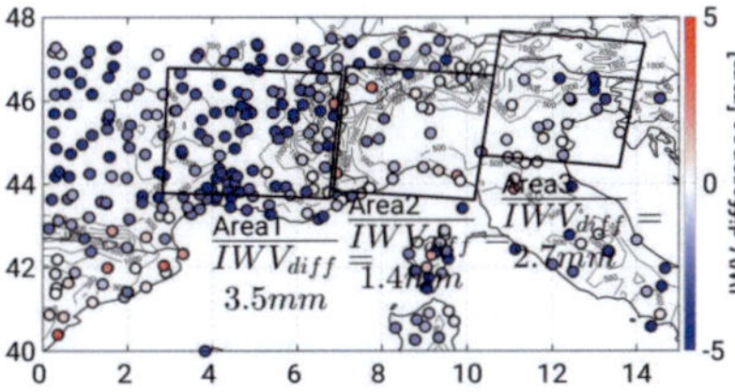

(c) CTRL-2.8

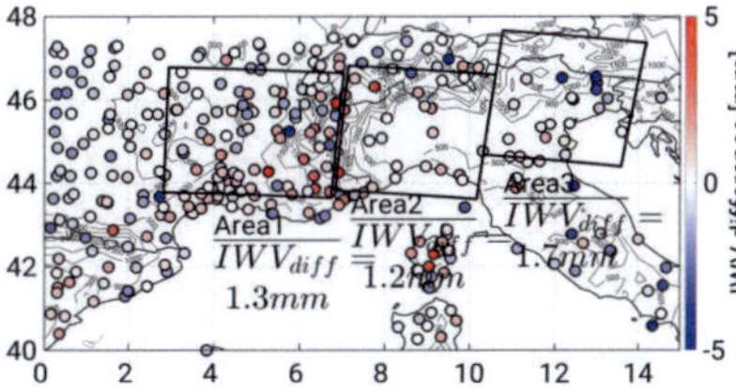

(d) AS-GPS-RAD-2.8

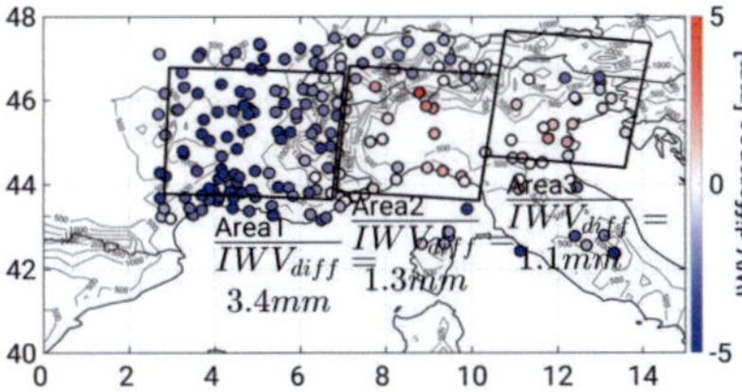

(e) CTRL-500

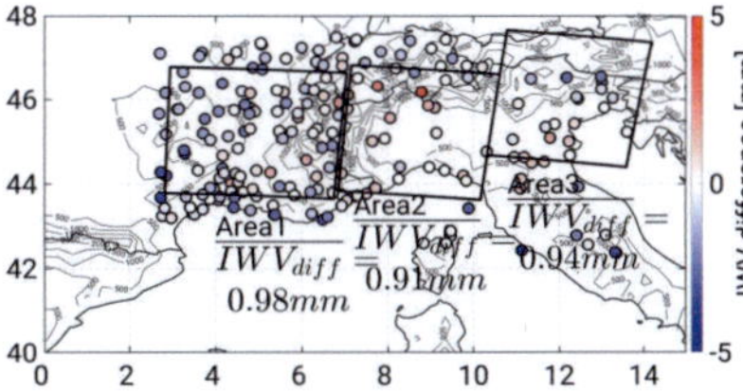

(f) AS-GPS-RAD-500

Figure 8.1: Spatial distribution of the differences in IWV between GPS and COSMO for CTRL-7 (a), AS-GPS-RAD-7 (b), CTRL-2.8 (c), AS-GPS-RAD-2.8 (d), CTRL-500 (e) and AS-GPS-RAD-500 (f). The differences are obtained as (OBS-COSMO) after averaging IWV for every GPS station during the preconditioning period i.e. 23-Sep 12:00h to 23-Sep 19:00h. The IWV given by COSMO is obtained by natural interpolation from the neighbouring grid points. Blueish colours stand for an IWV excess in COSMO as compared to GPS.

Table 8.1: (a) Mean Absolute Error (MAE) of spatially averaged hourly IWV between COSMO and GPS-IWV for the investigation areas and all simulations in the period 23-Sep 12:00h to 23-Sep 19:00h.

Mean abs. error	**7 km**			**2.8 km**			**500 m**		
[mm]	Area1	Area2	Area3	Area1	Area2	Area3	Area1	Area2	Area3
CTRL	2.4	2.2	2	3.5	1.4	2.7	3.4	1.3	1.1
AS-GPS	1.7	2	1.7	1.2	1.1	1.6	1.1	1.1	1
AS-RAD	2.6	3.6	2.8	1.8	2.4	2.4	1.5	1.4	1.3
AS-GPS-RAD	1.8	2.2	2	1.3	1.2	1.7	0.98	0.91	0.94

(b) Areal mean of daily accumulated precipitation on the 24-Sep-2012, over the three investigation areas (Area1, Area2 and Area3) as given by the different simulations.

$\overline{Precipitation}\,[mm\,d^{-1}]$	**7 km**			**2.8 km**			**500 m**		
	A1	A2	A3	A1	A2	A3	A1	A2	A3
CTRL	15.1	13.6	14	15.7	11.3	11.5	15.1	10.5	8.8
AS-GPS	11.9	9.1	8.8	14.5	9.5	10	12.7	7.9	6.1
AS-RAD	22.5	19	10	23	17.8	7.11	19.3	15	8.7
AS-GPS-RAD	16.3	11.5	9.12	20.6	13.4	8.2	16.2	12.3	8.8

The spatial distribution of the differences reveals a too wet atmosphere west of the Alps (southern France), a too dry atmosphere over northern Italy and strong positive deviations over elevated terrain (the Alps, Corsica and and central Italy). The latter two impacts are more apparent in the 7 km than in the 2.8 km runs and in 500 m. These impacts are in agreement with the results of Chapter 6.

Figure 8.1 highlights the good correction of the biases when applying the nudging. In particular, of the nudging of column atmospheric moisture, as expected. This holds for both the spatial distribution and the temporal evolution of IWV (see Fig. A.6, Fig. A.7 and Fig. A.8 in the Annex) and for all horizontal resolution types. Table 8.2a, summarizes the Mean Absolute Error (MAE) values for all simulations during this period and over the three investigation areas. The MAE is lowest for the AS-GPS and AS-GPS-RAD runs (values in the range 0.91 mm to 2.2 mm. On the contrary, the different AS-RAD simulations show, over some areas, larger biases than their respective CTRL counterparts. This is evidence of the added value of combining column atmospheric moisture measurements with radiosondes in the representation of IWV.

Regarding the comparison between the different grid spacings after nudging column atmospheric moisture, it is shown that the MAE improves as the grid length of model is refined, with best values shown by the 500 m grid for all AS runs. This can be explained, again, by the improved representation of the model surface height in the finer resolution grids.

The accuracy of the representation of the spatio-temporal distribution of humidity during IOP6 is further assessed, this time with respect to the vertical distribution. Figure 8.2 shows the Bias and RMSE of specific humidity between the COSMO simulations and the operational soundings of at Nimes (see location of the station in Fig. 5.4) in the 22-Sep-2012 to 25-Sep-2012 period. The radiosonde data has been interpolated to eleven COSMO pressure levels and the closest COSMO grid point to the location of the station is selected.

The top panels of Fig. 8.2 show the mean bias of specific humidity at Nimes for all simulations. Chap. 6.2, showed how COSMO-CLM overestimated specific

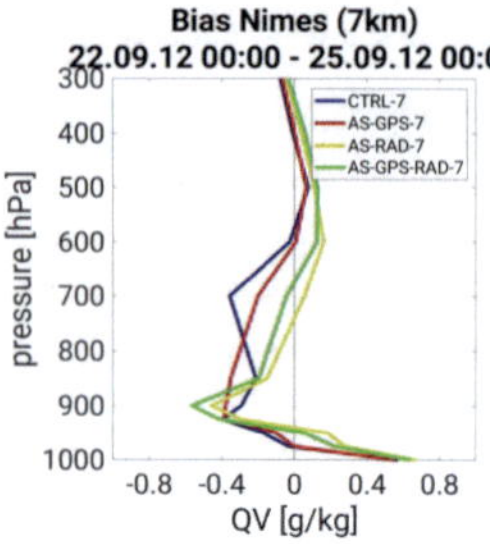

(a) Bias (7 km)

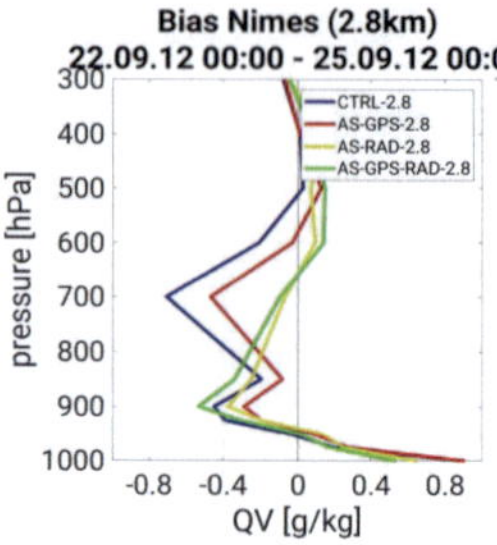

(b) Bias (2.8 km)

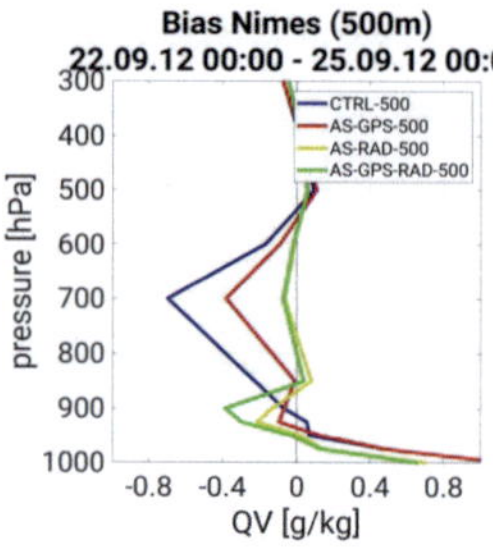

(c) Bias (500 m)

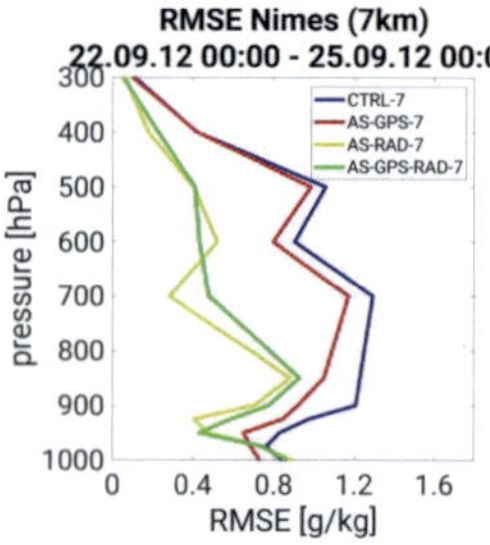

(d) RMSE (7 km)

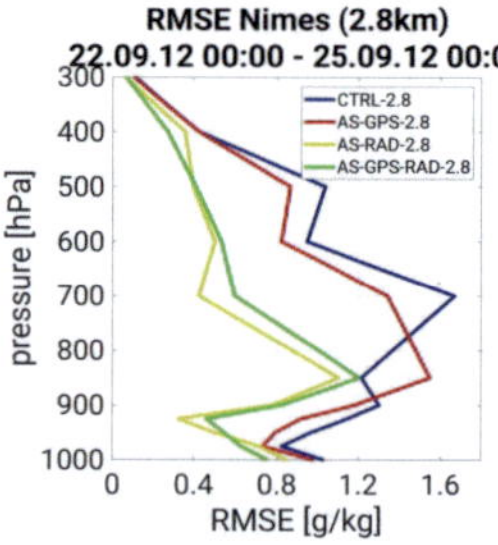

(e) RMSE (2.8 km)

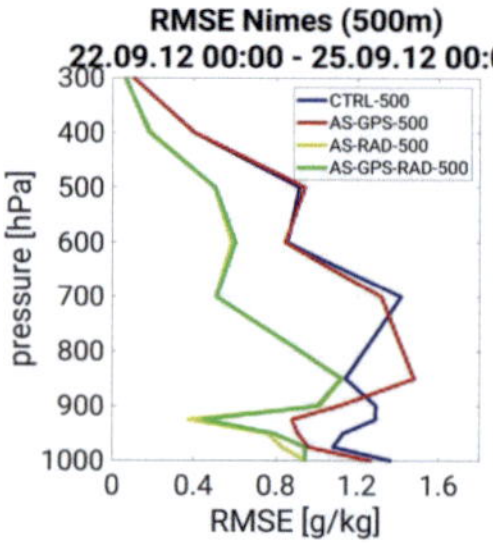

(f) RMSE (500 m)

Figure 8.2: Bias (top) and RMSE (bottom) of specific humidity in $g\ kg^{-1}$ averaged in time at the available pressure levels for all soundings at Nimes between 22-Sep 00:00h and 25-Sep 00:00h. The sounding data has been interpolated vertically to the eleven pressure model levels of COSMO. The COSMO grid point closest to the station location is used for the comparison

humidity above 900 hPa and underestimated it below for three selected low-elevation stations (Barcelona, Nimes and Pratica) in the seasonal simulations. The same issue occurs in the simulation of the vertical distribution of specific humidity during IOP6 at Nimes.

The nudging of the column atmospheric moisture and the soundings is most beneficial to dry the upper levels and bring better scores of the bias up to 500 hPa (for example 0.1 $g\ kg^{-1}$ at 700 hPa in the AS-RAD and AS-GPS-RAD for both grid length types). Nudging column atmospheric moisture observations, reduces slightly RMSE below 900 hPa (by about 0.2 $g\ kg^{-1}$ depending on the grid length type), and nudging radiosonde variables brings a clear improvement only when utilized in very fine grid lengths. This means that the combination of nudging sounding data and using very fine model resolutions is advantageous to the representation of the specific humidity profile. Nudging column atmospheric humidity and radiosonde data, helps improve the representation of humidity at the higher levels but this effect is not clear close to the ground. These conclusions highlight the need for improvement of the atmospheric moisture representation at the lowest atmospheric levels (below 900 hPa).

Precipitation is largely affected by the column moisture and radiosonde nudging during IOP6. Figure 8.3 shows the daily accumulated precipitation on the 24-Sep-2012 for the runs without any observation nudging (CTRL), the nudging of GPS data alone. i.e. column atmospheric moisture (AS-GPS) and the nudging with the soundings and the GPS data (AS-GPS-RAD) for the 7 km, 2.8 km and 500 m horizontal resolutions. All CTRL runs show the four main heavy precipitation spots (The Massif Central and Avignon regions, the Alps at the Swiss-Austrian region, the Italian-Slovenian border and the gulf of Genoa). Remarkable differences between the CTRL runs of all resolution types exist, specially in the structure of precipitation between the 7 km (convection parameterized) and 2.8 km and 500 m (convection permitting) grids. In agreement with past literature (Kain et al., 2008; Zhang and Zhang, 2012) reducing the grid size reduces the precipitation structures, with generally an intensification of precipitation. Besides, over Area3, both the 2.8 km and the 500 m do not represent such an intense precipitation at the Italian-Slovenian border in comparison with CTRL-7. Instead, they split this precipitation maximum between the mountains and the coast of Italy. This decrease in precipitation over A3 can be seen in Table 8.2b which shows the mean daily accumulated precipitation over the three investigation areas.

Overall, the impact of the column moisture nudging was to reduce precipitation intensity over all investigation areas. Besides, the column moisture nudg-

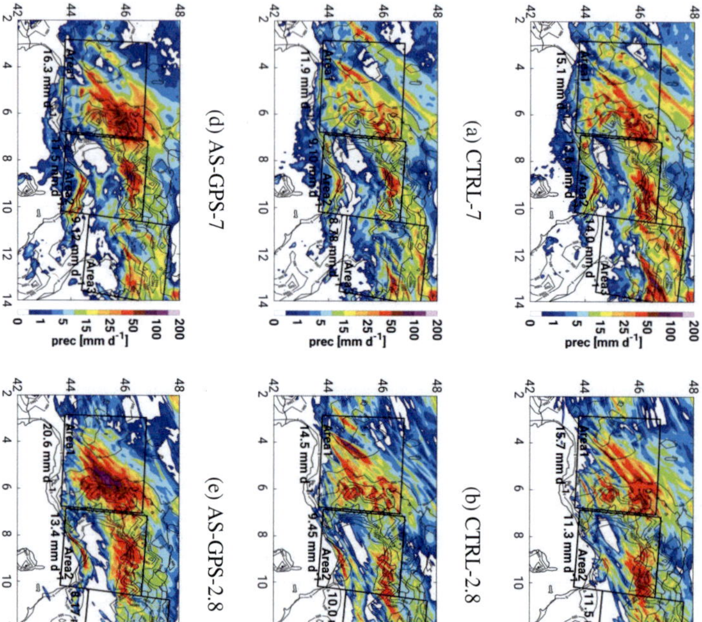

Figure 8.3: Spatial distribution of daily accumulated precipitation during 24-Sep-2012. The rows show the CTRL, AS-GPS and AS-GPS-RAD nudging type and the columns the 7 km, 2.8 km and 500 m horizontal grid resolution types.

ing seems not to modify largely the structure of precipitation and there are indications that the changes in humidity induced by the column atmospheric moisture nudging constrained its intensity but not in its location or structure. This impact is analogous for all horizontal resolutions (see figure 8.3).

Regarding the impact of the radiosondes nudging, Area1 and Area2 underwent an intensification of precipitation, and a spreading of the precipitation structure. Area3 showed a similar impact (intensification and spreading) but only in the 500 m runs. Mean precipitation was increased between 4 $mm\ h^{-1}$ and 7 $mm\ h^{-1}$ over Area1 and Area2 as a result of the radiosonde nudging for all grid resolution types. The more extensive precipitation structures can be explained considering the location of the nudged soundings and the horizontal spreading of the nudged data. The sounding stations that fall within investigation areas Area1 and Area2 are Nimes and Payerne (Area1) and Milano (Area2). The nudging of the measurements performed during the 24-Sep-2012 at these stations have a remarkable impact based on how precipitation maxima span grid points downstream of these locations. Especially impacting are the reports from Nimes (Area1) and Milano (Area2). The nudging procedure spread the information of the observations to neighbouring grid points by using a spatial weight which has a Gaussian decaying form (see Chap. 4.2).

When nudging column atmospheric moisture observations together with radiosondes, a combined effect on the precipitation structure and the intensity was observed. In other words, the nudging of soundings spread and increased precipitation and the column moisture nudging topped the impact of that increase. This regulation of maximum precipitation exerted by the column atmospheric moisture observations is perceptible in the comparison of the spatial distributions of the AS-RAD with AS-GPS-RAD runs (see figure 8.3). Thanks to the regulation in IWV brought by the column moisture nudging, the increases in the AS-GPS-RAD runs over Area1 and Area2 ranged between 1 $mm\ d^{-1}$ and 4.9 $mm\ d^{-1}$ when compared against CTRL for all resolutions.

To assess the accuracy of the different simulations of precipitation, an object based verification method known as the Structure Location Amount (SAL)

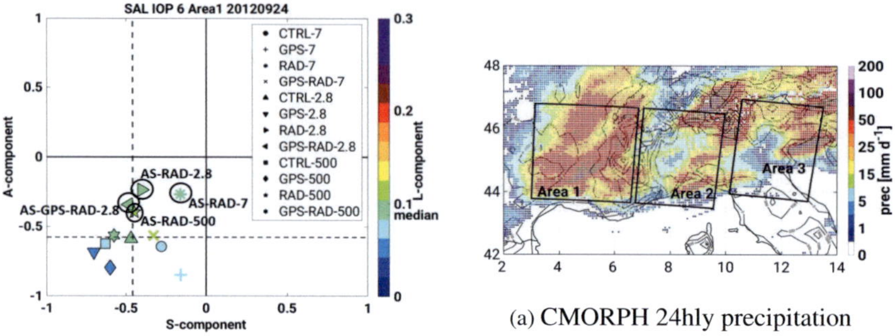

(a) CMORPH 24hly precipitation

Figure 8.4: Structure Location Amount (SAL) diagram for the COSMO vs CMORPH comparison. All simulations are shown and all grids have been upscaled to the CMORPH (8km) grid. The CMORPH daily accumulated precipitation fields (right panel) are compared to the daily accumulated precipitation over Area1 (see figure 8.3). The SAL metrics close to zero indicate a good agreement of COSMO against the observations.

method (see chapter 5) is presented. Figure 8.4 and Table A.1 of the Annex, shows the values of the three metrics, presented for comparison of the different COSMO simulations against the remote sensing precipitation measurements from CMORPH (see chapter 5 for a description of the data set). The compared precipitation fields, are the daily accumulated precipitation as represented by COSMO (figures 8.3) and by the CMORPH data set (figure 8.4a).

The three SAL metrics show that the best performance in simulating heavy precipitation over Area1 and Area2 is given by the AS-RAD or the AS-GPS-RAD runs for this HPE (see Table A.1 in the Annex). This holds for all grid length types. Area3 shows a different response to the nudging of the observations where not a clear improvement by the sounding can be found and, for instance, the CTRL-500 or the AS-GPS-2.8 runs showed the best SAL results. Area3 was shown to have very different impacts in the representation of precipitation among the different nudged simulations and grid-length types used. The structure and amount of precipitation were specially improved over Area1 and Area2 when the radiosondes were introduced. Over Area2, for example, all SAL metrics improved in the AS-GPS-RAD simulations obtained with the 2.8 km and the 500 m grid lengths (see Fig. 8.4). A better agreement with CMORPH observations is gained over Area1 and Area2 due to the enhanced precipitation along the Avignon valley with more spreading towards the western Flank of the Alps and over the Swiss Alps close to the Austrian border in Area2.

8.2 Mechanisms Inducing Precipitation Intensification in 500 m Simulations

The analysis of the impact of radiosonde data in the nudging procedure has revealed an interesting enhancement of precipitation accompanied by spreading of the structure over two investigation areas during IOP6. This effect, was observed in all simulations of all resolution types. In the following, the changes induced by the nudging in the processes leading to convection are investigated in detail to find the most relevant mechanisms, explaining the precipitation intensification. Area1 is selected in this section, for in-depth process-understanding. The panels shown in this section account for the runs in a 500 m grid, given the similar responses to the nudging in the 7 km and 2.8 km. Results obtained for these two grids can be found in the Annex. When existing, the relevant differences will be put forward.
Figure 8.5 shows the temporal evolution of hourly IWV (top) and precipitation (bottom), spatially averaged over Area1 obtained with the different simulations in a 500 m horizontal resolution grid. The figure shows two periods of large deviations between the CTRL and the rest of the simulations. These occur between 23-Sep 08:00h and 23-Sep 21:00h, with maximum differences in the spatial averages in the order of -3 *mm*, and between 24-Sep 04:00h and 24-Sep 09:00h, with differences up to $+3$ *mm*, for the AS-RAD-500 and AS-GPS-RAD-500 simulations against CTRL-500. The simulations using a 7 km and a 2.8 km grid showed the same periods of large differences albeit these lasted less in the 7 km case. Moreover, the AS-RAD and AS-GPS-RAD runs obtained

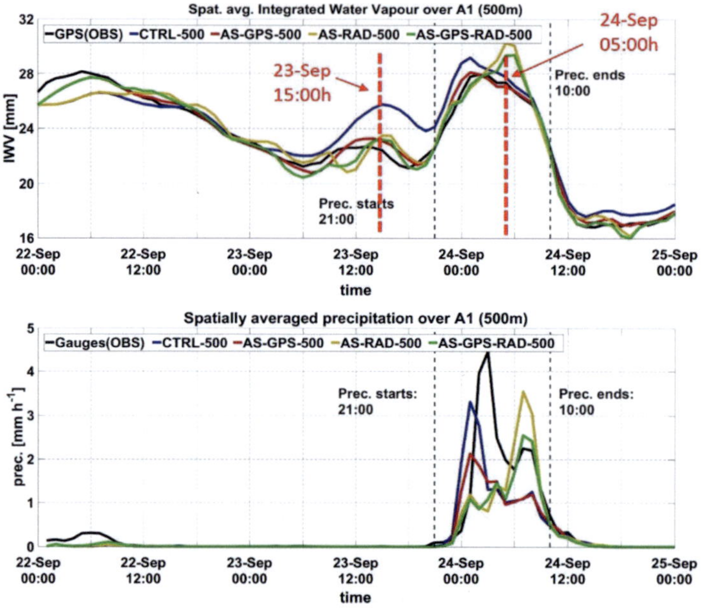

Figure 8.5: Spatially averaged IWV (top) and precipitation (bottom) over Area1 (see Fig. 5.4, for the location of the investigation area) for the COSMO runs on the 500 m grid (colours), GPS stations and Rain Gauges (black). The spatial averages are done selecting the grid-points closest to the stations, neglecting those further than 1km away.

with the 7 km and the 2.8 km horizontal resolutions showed larger errors (of about -2 *mm*) during the first of the two pointed out periods, i.e. during the preconditioning.

The occurrence of these two periods of a strong humidity modification explain the evolution of precipitation over Area1. As shown in the lower panel of figure 8.5, COSMO represents a first precipitation maximum in the CTRL-500 and AS-GPS-500 runs (red and blue lines), at 24-Sep 01:00h followed by a later maximum in the AS-RAD-500 and AS-GPS-RAD-500 simulations (yellow and green lines), at 24-Sep 08:00h. Differences in spatially averaged precipitation intensity are of -1 $mm\ h^{-1}$ between CTRL-500 and AS-GPS-500 during the first period and of about 2 $mm\ h^{-1}$ between CTRL-500 and AS-

RAD-500 and AS-GPS-RAD-500 during the second period. These precipitation differences can be related to the IWV differences assessed in the previous paragraph. The second precipitation maximum is shown by the AS-RAD and AS-GPS-RAD runs of all resolutions albeit being weaker in the 7 km horizontal resolution simulations. The first maximum comes from precipitation occurring first over the Massif Central mountain system (west of the investigation area, at about 24-Sep 02:00h) and the second over the western flank of the Alps (east side of the investigation area; at about 24-Sep 08:00h). As shown in section 7.4, the synoptic and mesoscale conditions present at the region evidenced a persistent low-level moist inflow from the Mediterranean starting on the 21-Sep-2012 and instability building up along the Avignon valley. Therefore, the differences in precipitation of the first maximum stem from the modification of instability and moisture amount over the area in the nudged runs prior to the triggering of DMC (preconditioning). Conversely, the second precipitation maximum of the AS-RAD and AS-GPS-RAD simulations during the second maximum depends on the conditions present over Area1 once precipitation had already started.

To investigate the role of preconditioning during the first precipitation maximum, the conditions over Area1 between 23-Sep 12:00h and 23-Sep 19:00h are summarized in figure 8.6 using box and whiskers diagrams of some atmospheric variables. They show, that the triggered convection at the arrival of the mesoscale front is preconditioned by intense moisture flux at 700 hPa (especially in the CTRL-500 run) and a favourable CAPE-CIN balance.

When convection is triggered, latent and potential instability were largest for CTRL-500, AS-RAD-500 and AS-GPS-RAD-500. Additionally, CIN played a role in hindering the probability of lifted air parcels reaching the Level of Free Convection (LFC) for all nudged (AS) runs. Finally, the lower free-troposphere was dried in all nudged (AS) runs, especially those containing radiosonde data (AS-RAD-500 and AS-GPS-RAD-500), as shown by the moisture flux at 700 hPa. These impacts explain the reduced precipitation in the nudged runs compared to CTRL. As assessed in past studies (Takayabu et al., 2010; Hohenegger

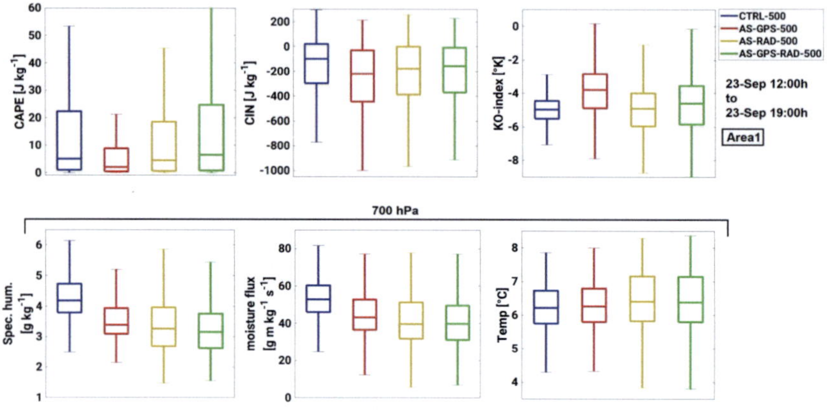

Figure 8.6: Box and whiskers plots showing the median (mid horizontal line), the percentiles 25 and 75 (the box edges) and the extreme values (whiskers) of the different simulations using a 500 m grid resolution. All box and whiskers are obtained from the distribution of values for the shown quantities between the 23-Sep 12:00h and 23-Sep 19:00h. The top panel shows the distribution of instability indices (CAPE, CIN and KO-index) and the bottom panel shows specific humidity, moisture flux and temperature at 700 hPa.

and Stevens, 2013) a drying of the the environment at 700 hPa favours convection termination by dry-air entrainment in the cloud plumes.

To further investigate the convective preconditions present at Area1 simulated by COSMO, Figure 8.7 shows the longitude-height cross sections of Equivalent Potential Temperature (θ_e) in colour shading, specific humidity, in black contour lines, and meridional and vertical winds for CTRL-500 and AS-GPS-RAD-500. The shown hour is 23-Sep 15:00h, i.e. the moment when the difference in IWV between CTRL and the nudged runs of all grid resolutions are largest in the preconditioning, 8h prior to precipitation. The cross-section is selected at a longitude of $45.75°N$ as shown in figure 5.4.

In agreement with the box and whiskers plots of figure 8.6 the vertical distribution of specific humidity shows a very moist PBL. There, values as large as $12\ g\ kg^{-1}$ are represented over the western part of the Massif Central and the western foothills of the Alps. Large specific humidity values are also found at

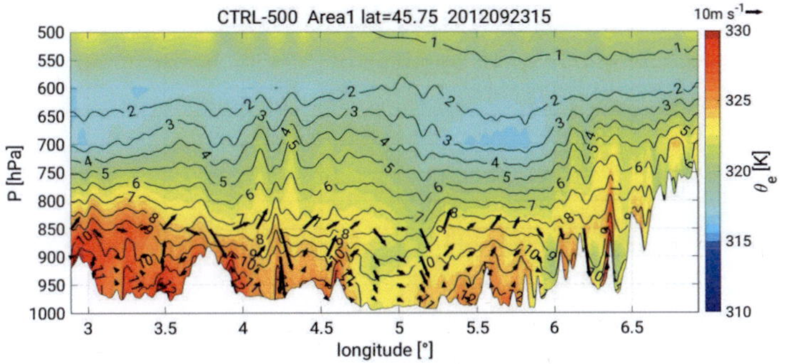

(a) CTRL-500

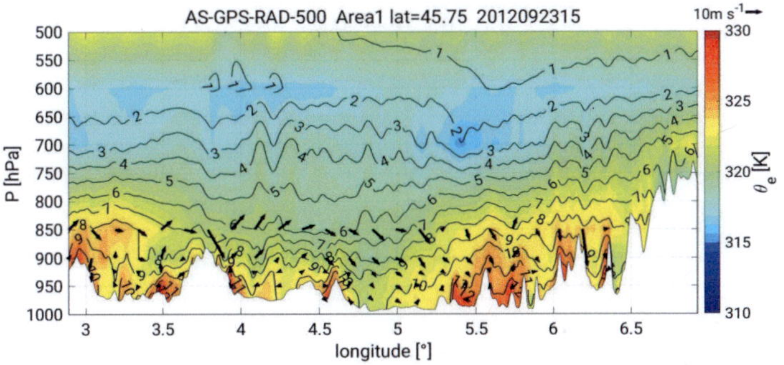

(b) AS-GPS-RAD-500

Figure 8.7: Cross-section at latitude 45.75° of Equivalent Potential Temperature (θ_e) in colour shading, zonal-vertical winds, and specific humidity (black contours) on the 23-Sep 15:00h for CTRL-500 (top) and AS-GPS-RAD-500 (bottom). The orography is shown in white.

the PBL in the CTRL-7 and CTRL-2.8 runs albeit these were somewhat lower (10 $g\ kg^{-1}$ in CTRL-7 and 11 $g\ kg^{-1}$ in CTRL-2.8). This is due to the inability of the coarser resolution grids to resolve small scale specific humidity variations (of the order of 1km) present for example at 3.5°E in the CTRL-500. Humidity at the PBL over the Massif central is lowered in all nudging types by about 1 $g\ kg^{-1}$ but not so remarkably at the Alpine foothills, this impact is similar for 7 km and 2.8 km.

Regarding θ_e, a large gradient is found along the cross-section with a change of ca. 15 K between the 950 hPa and 650 hPa levels in the CTRL-500 run. The lower θ_e at 650 hPa is decreased further by the sounding nudging i.e. in the AS-RAD-500 (not shown, values of ca. 310 K) and in AS-GPS-RAD to ca. 312 K. The reduction is weakened in the latter given the regulation of the humidity variation by the column moisture nudging. This effect is weaker as the grid resolution is increased and, indeed, the extension of this area of low equivalent potential temperature at 650 hPa spans ca. 0.5° in the 500 m runs but only 0.01° in 2.8 km and disappears in AS-RAD-7.

When precipitation is already occurring over Area1 an abrupt precipitation increase is shown by the AS-RAD and AS-GPS-RAD runs of all resolution types at about 24-Sep 08:00h. The nudging of the 05:00h Nimes radiosonde forces Area1 and its surroundings toward saturation between 600 hPa and 950 hPa. Figure 8.8 shows the vertical distribution of dew point temperature (dashed lines) and temperature (straight lines) at that hour for the simulations in the 500 m grid. An analogue impact of the nudging of the 05:00h sounding is shown in the 7 km and 2.8 km grids. This impact lasts for about 3h and ex-

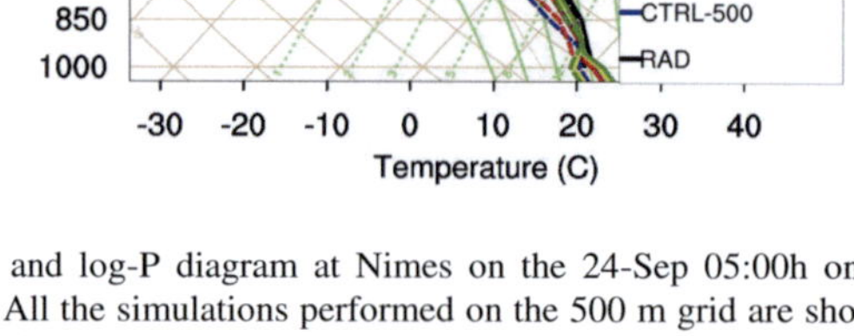

Figure 8.8: Skew-T and log-P diagram at Nimes on the 24-Sep 05:00h once precipitation has started. All the simulations performed on the 500 m grid are shown (colours) as well as the nudged sounding (black). The grid points closest to the sounding are shown.

plains the more intense and spread precipitation over the western flank of the Alps within Area1 (see Fig. 8.3).

The nudging of the 05:00h radiosonde at Nimes forced saturation, bringing a deeper convection together with less dry air entrainment in agreement with a more moist environment present up to to the mid troposphere. Figure 8.9 aids quantifying the impacts on specific humidity, wind speed, moisture flux, temperature and vertical winds at the 500 hPa level. The results are shown for the 500 m grid but the impacts are analogous to the 7 km and 2.8 km grids. The box and whiskers plots show the median, the percentiles 25 and 75 and extremes of the period 24-Sep 05:00h to 24-Sep 10:00h, i.e. the period of influence of the 05:00h sounding nudging.

Between 500 hPa (see Fig. 8.9) and 700 hPa (not shown), temperature, specific humidity and moisture flux show differences as large as 2 $°C$, 0.25 $g\ kg^{-1}$

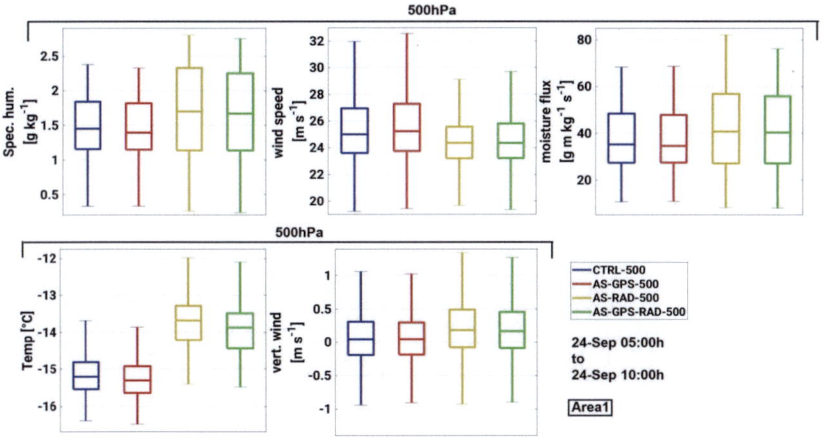

Figure 8.9: Box and whiskers plots showing the median (mid horizontal line), the percentiles 25 and 75 (the box edges) and the extreme values (whiskers) of the different simulations using a 500 m grid resolution. All box and whiskers are obtained from the distribution of values for the shown quantities between the 24-Sep 05:00h and 24-Sep 10:00h. The top panel shows the distribution of specific humidity, wind speed and the moisture flux at 500 hPa. The bottom panel shows the temperature and vertical at the same height.

and 10 $g\ m\ kg^{-1}s^{-1}$ respectively, between the nudged runs with radiosondes and CTRL and AS-GPS. This is in agreement with the skew-T log-P profile representation of Figure 8.8 and implies deeper saturation and a moister environment impeding dry air entrainment at this layer. This is in spite of the fact that horizontal wind speed is somewhat more intense for CTRL-500 and AS-GPS-500 runs than their AS-RAD and AS-GPS-RAD counterparts (winds speeds are larger by ca. 2 $m\ s^{-1}$). Indicating a large contribution to the moisture flux by the specific humidity. All in all, the AS-RAD and AS-GPS-RAD simulations are able to represent stronger convective updrafts (extreme positive values ca. 0.5 $m\ s^{-1}$ larger than CTRL and AS-GPS) which explain the more intense precipitation over the area.

Conclusions

Q.1 *Can added value be gained in the precipitation and humidity representation of HyMeX-IOP6 by nudging highly sampled humidity observations (GPS-ZTD), jointly with realistic atmospheric profiles and refining the model grid (7 km, 2.8 km and 500 m)?*

The combined use of very fine grid horizontal resolutions and the nudging of column atmospheric moisture and radiosondes improved the representation of the Root Mean Square Error (RMSE), the Correlation Coefficient (R) and the Agreement Index (AI) of IWV in its spatially averaged temporal evolution (not shown) and the Mean Absolute Error (MAE), in its spatial distribution (see Figure 8.1 and Table 8.2a). The nudging of column moisture observations with a 10 minute temporal resolution alone, improved the MAE of IWV for all different horizontal resolutions and the improvement is larger as finer horizontal resolutions are utilized. The nudging of radiosondes alone (every ca. 6h), did not improve the representation of IWV. However, the MAE scores were better for the AS-RAD simulations in the 500 m and the 2.8 km resolutions (see table 8.2a) as compared to 7 km, indicating added value of nudging radiosondes on model grids with larger numbers of vertical levels and finer horizontal resolutions. Finally, nudging both types of observations together was shown to be most beneficial for the 500 m and 2.8 km runs.

The areal RMSE, R, AI or MAE of IWV between COSMO and GPS did not show a systematic improvement due to the refinement of the model grid, alone, for the different investigation areas (see table 8.2a). Still, refining the model grid, was shown to correct the bias of particular stations located over steep terrain (see Fig. 8.1) whose height difference against the model surface was shown to be too large.

Regarding the vertical distribution of humidity, the assessment on the impact of the nudging during this event over southern France (Area1) showed a slight improvement brought by column moisture nudging, alone (by drying the levels between 500 hPa and 900 hPa) and a large improvement by the the runs including radiosonde data at those levels (see Fig. 8.8). Below 900 hPa likewise the SOP1 seasonal simulation with COSMO-CLM, the nudging of column moisture observations alone could not correct the underestimation of specific humidity but improves the scores close to the surface. Still biases up to 0.6 $g\ kg^{-1}$ perist (see Fig. 8.2).

The representation of precipitation was improved by the inclusion of the radiosondes for all resolution types and for investigation Area1 and Area2, as demonstrated by the SAL verification (see Table A.1). The best SAL values in the comparison against CMORPH daily precipitation, were given either by the AS-RAD or the AS-GPS-RAD runs for all resolution types. The improvement in the representation of precipitation was brought about an intensification and spreading of precipitation over the east of Area1 and the northern part of Area2. This intensification and spreading occurred after the nudging of the radiosondes obtained at Nimes in the 24-Sep 05:00h and at Milano at 05:00h and 11:00h.

Nudging radiosondes, enhanced precipitation over Area1 and Area2 for all resolutions. For its part, the column moisture nudging, did not affect much the structure or the location of the maximum precipitation but it regulated the amount. The combined use of GPS and radiosondes was most beneficial for the 7 km and 500 m runs over area Area1 and for the 2.8 km over Area3.

Q.2 *What is the impact for the processes leading to convection of nudging column atmospheric moisture observations combined with atmospheric soundings in very high-resolution simulations of IOP6?*

A strong impact by the nudging of column atmospheric moisture observations and radiosondes on convective-related processes was assessed in the representation of Deep Moist Convection (DMC) evolution. Two phases of the impact of the nudging were found for this event at southern France (Area1). These were, the preconditioning by moistening and instability build-up, between 23-Sep 12:00h and 19:00h, and the deep saturation forcing by the nudging of the 05:00h Nimes radiosonde, between 24-Sep 05:00h and 10:00h (see Fig. 8.5). Both phases are present, in the simulations with a 7 km, 2.8 km and 500 m grid horizontal resolutions, yet with minor differences in timing, extent and fluctuations. The impact of the nudging assessed was analogous to all resolution types for Area1 and for this reason the results on the 500 m grid are selected for quantification.

The nudging of column moisture and radiosondes during preconditioning (in this HPE, a persistent moistening of Area1 more than two days prior to precipitation) induced a humidity reduction (ca. 4 mm less) in all AS runs, 12h prior to the initiation of precipitation. This drying, topped the instability of the atmospheric column over Area1 especially for AS-GPS-500 (changes in extreme KO-index of $+3\ ^{\circ}C$ and in CIN of $+100\ J\ Kg^{-1}$), and for AS-RAD-500. Besides, this drying is responsible for a weaker moisture flux most remarkably at 700 hPa , in the AS-GPS-500 and AS-GPS-RAD-500 (see Fig. 8.6) with decreases of ca 20 $g\ m\ kg^{-1}\ s^{-1}$. Although to a lesser extent, this moisture flux reduction during preconditioning is also apparent in the AS-RAD-500 run (decrease of ca. 15 $g\ m\ kg^{-1}\ s^{-1}$). These results show that the most relevant process hindering convective precipitation during the first 6h of the event in the AS runs was the entrainment of dry air at the lower free-troposphere reducing the buoyancy of the plumes. The analysis on the cross-sections at a particular latitude (N45.75°) of Equivalent Potential Temperature (θ_e) and Specific Humidity (q_v) revealed a deep very moist layer, from the surface up to 800 hPa in the CTRL-500 run with ca. 10 $g\ kg^{-1}$, and a large θ_e gradient of ca. 15$°K$ between the surface and 600 hPa , which is reduced in the AS runs (see Fig. 8.7).

Regarding the period of strong saturation forcing by the nudging of the Nimes radiosonde at 05:00h, the moisture flux was, again, the variable whose impact showed the largest relevance in the evolution of precipitation during the last 5h of the event. The layer between 500 hPa and 850 hPa underwent a large moisture flux increase ($+10\ g\ m\ kg^{-1}s^{-1}$) and temperature increase ($+2°C$) in the AS-RAD-500 and AS-GPS-RAD-500 runs. The latter impact on temperature is explained by the deeper saturation of the atmospheric column over Area1, as demonstrated by the SkewT-logP plots (see Fig. 8.8) and the release of latent heat flux in the condensation process. This effect was present in the runs with radiosonde nudging for about 3h and explains the intensification of precipitation addressed in Figure 8.3. Given the horizontal spreading of the nudged vertical profile, the precipitation structures spread along the western side of Area1.

9 Conclusions

The prevention of the impacts related to heavy precipitation is of uttermost importance to reduce the monetary losses and casualties that every year occur in the western Mediterranean (WMed) region. Under climate change the WMed region is projected to endure this high impact weather more frequently and hence, accurate forecasting is crucial for mitigating its effects and supporting the national weather services and emergency units. Real benefit to society can be brought through the improvement of the Numerical Weather Prediction (NWP) systems and Climate models which still show deficiencies in representing phenomena responsible for heavy precipitation such as Deep Moist Convection (DMC). Deficiencies in simulating DMC arise partly from the misrepresentation of the spatio-temporal distribution of atmospheric moisture in the models as well as of the manifold processes by which it influences atmospheric convection. Advancements in this regard may only be achieved through a deep understanding and improved representation of the physical phenomena behind these events.

This thesis aims at gaining knowledge and reducing model uncertainty associated with the interaction of atmospheric moisture and convection, assessing the impact of nudging observations of column atmospheric moisture in different resolution simulations from seasonal to event scales. To this end, atmospheric moisture sensitivity experiments have been carried out by nudging Global Positioning System (GPS)-derived Zenith Total Delay (ZTD) observations, a proxy for column atmospheric moisture, with a temporal resolution of 10 minutes and a large coverage during a complete autumn season in the WMed. The Consortium for Small-scale Modelling (COSMO) model is utilized in horizontal grid-spacings of 7 km and 2.8 km in order to assess the sensitivity to the moisture nudging by convection-parameterized, as well as, by convection-explicit simulations. Moreover, very high resolution simulations (500 m) have been performed to analyze the 24-Sep-2012 heavy precipitation case study, for the first time. Novel quantifications, analysis and impacts of the moisture nudging experiments are presented for selected regional and local investigation areas with semi-arid and humid Mediterranean Climate systems. The results presented in this thesis support in any case the relevant role of atmospheric moisture in convective precipitation. The main findings are summarized in the following:

Column moisture nudging impact on water vapour distribution

- Nudging column atmospheric moisture is of great benefit to correct the humidity wet bias present in the simulation of the autumn 2012 season in the WMed with COSMO. This was demonstrated by the improvement in the Root Mean Square Error (RMSE), the agreement index and mean absolute error in the 7 km and 2.8 km horizontal resolutions using the climate and the weather versions of COSMO. The largest corrections took place over the Iberian Peninsula(IP) and France (FR) for both horizontal resolutions (RMSE reduced from ca. 1.8mm to 0.4mm) and the shortest over IT (RMSE reduced from 1.9mm to 0.8mm). This was the case given IT has less GPS stations and is affected typically by southwesterly flows which cross over regions with no GPS coverage (Africa, the Mediterranean). A humidity wet bias in the COSMO model had already been reported by past studies (Schraff et al., 2008; Cress et al., 2012; Devidasrao, 2012). The autumn 2012 wet bias stemmed in part from four sub-periods of large atmospheric moisture advection overestimated by COSMO (see Sec. 6.1). These sub-periods of two to three days were validated against the Moderate Resolution Imaging Spectroradiometer

(MODIS) instrument and against daily Integrated Water vapour (IWV) measured with GPS receivers. Three of the highlighted sub-periods were related to intense evapotranspiration taking place over the Iberian Peninsula and northern Africa, reaching the European investigation areas 1 to 3 days later.

- COSMO underestimates the column atmospheric moisture (ca. 6mm) in the seasonal and the event simulations at particular stations over elevated terrain (e. g. the Alps, the Pyrenees). These errors stem from a misrepresentation of the ground height. Given nearly 50 % of the total atmospheric moisture is contained within the first 1.5 km (Munn et al., 2003), the misrepresentation of the ground height induces large errors which could not be overcome by nudging column atmospheric moisture observations. This was due, first, to the fact that the nudging procedure dismissed the observations at these stations due to the height difference. Second, corrections induced by remote moisture observations were insufficient. In this sense using finer horizontal resolutions (2.8 km and 500 m) was useful given their improved representation of the model height.
- The vertical distribution of specific humidity in COSMO was shown to suffer from overestimation above the 900 hPa level (mean bias of about $-0.4\ g\ kg^{-1}$) and underestimation below (mean bias of $0.8\ g\ kg^{-1}$), at three selected stations over low terrain (Barcelona, Nimes and Pratica di Mare). This was the case for the seasonal and event scale simulations in the climate and the weather modes and for the three used horizontal resolutions (7 km, 2.8 km and 500 m). These deficiencies in the vertical distribution arise from the inability of COSMO to represent the steepness of the real humidity vertical gradient. Nudging column atmospheric moisture observations reduced the overestimation above 900 hPa (mean bias of $0.2\ g\ kg^{-1}$) reduce these errors. When combined with ra-

diosondes, improvement was gained also below 900 hPa (mean bias of 0.6 $g\ kg^{-1}$) especially in the 2.8 km and 500 m grids.

Moisture effects on the processes leading to convection and precipitation

- The correction of the humidity wet bias in the seasonal simulations, brought a reduction in IWV and specific humidity between -8 % and -3 %, depending on the study region, horizontal resolution or height considered. Larger IWV variations took place over France (FR) and the Iberian Peninsula (IP), about -8 %, than over IT (about -4 %) for the 7 km and 2.8 km resolution simulations. As a result, the seasonal mean conditions showed an overall decrease in 2m specific humidity (ca. $-0.3\ g\ kg^{-1}$), instability as given by CAPE (ca. $-10\ J\ kg^{-1}$) and of the total cloud cover (ca -4%). Most remarkably for low and medium clouds. All in all these impacts reduced the total precipitation amount over land (with decreases ranging between -9 % and -65 % which were largest for FR and shortest for IT. Over IT they were only weakly affected by the changes in atmospheric moisture, provided there are no GPS observations over water bodies.

- The seasonally averaged latent and sensible heat fluxes and 2m-temperature showed a heterogeneous spatial distribution of the differences induced by the moisture nudging. COSMO compensated regions of enhanced (reduced) outbound latent heat flux, by reducing (enhancing) the outbound Sensible Heat flux. This in turn, reduced (increased) 2m-temperature at those same regions (see Tab. 7.4 and Tab. 7.7). The heterogeneous spatial distribution of the seasonal differences over the sea and elevated regions (>3000 m) can be explained by the surface overall humidity reduction. Over the sea and very high mountains, suite to a humidity reduction close to the surface outbound latent heat emission was enhanced. Over land, the nudged runs showed the enhanced outbound latent heat flux emissions only over areas with enhanced seasonal precipitation.

- The changes in humidity affected the distribution of the large-scale pressure. The seasonal means showed an increase of the probability of higher geopotential heights (see Fig. 7.14). This was observed for FR and the AS-7 simulation. The analysis of a heavy precipitation event showed a weakening of a low pressure system over FR and IP, affecting, in turn, the associated wind circulation (see Fig. 7.21). This weakening of low-pressure systems over FR was caused by the humidity reduction entailed by the moisture nudging.

- The processes leading to convection most affected by the column atmospheric moisture nudging were conditional and potential instability (CAPE and KO-index) and drying (moistening) of the lower free-troposphere affecting entrainment/detrainment processes. These impacts explain to a large extent the variations in convective precipitation. This holds for the seasonal (see Chap. 7.3) and event scale simulations (see Chap. 7.4) on the 7 km and 2.8 km horizontal resolutions.

- The column atmospheric moisture nudging brought a large impact in the precipitation amount, structure and location. Precipitation RMSE, the correlation coefficient and structure, as given by the Structure Location Amount (SAL) method, in the seasonal simulations, improved as a result of the moisture nudging. This was the case over all climate regions. However, the moisture nudging was unable to systematically improve the representation of heavy precipitation events in the period. The non-systematic improvement, highlights the need for in-depth process-understanding of the impacts considering the particular characteristics of the affected regions and the dominating atmospheric situations.

- Preconditioning of heavy precipitation by moistening has been quantified to happen in a time window of ca. 9 h in mean, having a magnitude of $+4$ *mm* in IWV. About 87 % of the identified events in the period were preceded by an IWV monotonic increase which was shown to be largest ($\Delta IWV > 10$ *mm*) and longest ($\Delta t \approx 30$ *h*) for the extreme precipitation events (i.e. events falling within the percentile-90). Finally, it was also shown that an IWV increase (decrease) induced by the column atmospheric moisture nudging precedes an increase (decrease) in heavy precipitation in ca. 60 % of the events. These new findings add more information to the results by Schiro et al. (2016) which state that specific humidity is largest in the ± 3 *h* lag-lead time window, but do not quantify how early the moisture picks up. They also supplement the conclusions by and Khodayar et al. (2016a) reporting a moisture increase up to $6-24$ *h* before the initiation stage by using a different approach. These quantitative results were obtained by means of the *IWV-precipitation relationship* method (see section 7.2), developed in the course of this thesis which use is encouraged with observational data in future water vapour investigations.

Combined nudging of column atmospheric moisture and radiosondes to improve the vertical humidity and precipitation representation

- Combining column atmospheric moisture observations with radiosondes in the nudging experiments brought the best results in the representation of IWV, specific humidity and precipitation for a selected heavy precipitation event. The improvement in the humidity representation was specially remarkable in the finer grid-spacings (2.8 km and 500 m). Nudging radiosonde, corrected model errors in the vertical distribution of specific humidity, in the order of 1 $g\ kg^{-1}$, which were not able to be overcome by the column atmospheric moisture nudging alone. The modification of the water vapour stratification forced a deeper saturation, bringing a more intense and spread precipitation structure. This impact and improvement took place in all horizontal resolution types (7 km , 2.8 km and 500 m). This highlights the synergy of combining profile information with column atmospheric moisture measurements. The combined use of these observations is encouraged, especially, the use of the newest sounding systems which can sample humidity at a speeds faster than 10*s* (every 50m).

Dynamical downscaling and the moisture nudging experiments

- The use of different model horizontal resolutions (7 km, 2.8 km and 500 m) has allowed the study of the moisture effects on convection in different configurations (see Chap. 5). The most relevant similarities and differences found for the reference runs (no nudging applied) are: similar seasonal and IWV and specific humidity representation (See Fig. 6.8 , Tab. 6.2 and Fig. 6.9); narrower and more intense precipitation and convective updrafts in 2.8 km and 500 than in 7 km, in agreement with past findings (Fosser et al., 2014; Prein et al., 2015; Barthlott and Hoose, 2015); under atmospheric moisture reductions, the Tiedtke scheme (7 km) triggers less convection due to insufficient moisture amount below the cloud base, this effect is not present in the 2.8 km runs (see Chap. 7.2). On the other hand the impact of the column moisture nudging showed the following similarities and differences depending on the used model resolution: the finer the model horizontal resolution (2.8 km and 500 m), the better the improvement brought in humidity by the nudging of observations (see Tab. 6.2, Fig. 8.2 and Fig. 8.1) and the improvement in the representation of the Probability Distribution Function (PDF) and diurnal cycle of precipitation is best for the simulation in a 2.8 km grid nudging column atmospheric moisture (see Fig. 7.7 and Fig. 7.7).

These conclusions add new explanations to the existing literature on: the humidity bias in a widely used model, the large sensitivity of extreme and seasonal precipitation representation to moisture changes, the timing and magnitude of IWV variations in preconditioning heavy precipitation and the improvement in humidity and precipitation representation of nudging GPS and radiosonde in sub-kilometer model simulations. The assessed model impacts and biases are encouraged to be taken into account by weather prediction system developers and users to improve the representation of the convective processes and heavy precipitation. These conclusions show that the use of column atmospheric moisture observations (GPS-ZTD) is of great benefit to correct the humidity spatio-temporal distribution biases with a large impact on precipitation. In the outlook, performing analogue sensitivity experiments with state-of-the art remote sensing humidity profile products, such as those obtained from AIRS, AMSU or GPS radio occultation techniques is encouraged.

A Annex

7km	CA	VA	BA	CV	CO	LT	CI	NEI
IOP1				25% weakly				
IOP2								-25% weakly
IOP4							25% strongly	
IOP6				-40% weakly		-50% weakly		
IOP7a				65% weakly				
IOP7b						-5% weakly		
IOP8		30% weakly	60% weakly	-67% strongly				
IOP12a				-70% strongly	-65% weakly	-20% strongly	-75% weakly	
IOP12b			-30% weakly					
IOP13				-35% strongly	5% strongly	-30% weakly	-10% weakly	0% strongly
IOP14				-35% strongly				
IOP15a	-70% weakly	25% strongly	-10% weakly					
IOP15b				-50% strongly				
IOP15c					20% strongly			
IOP16a	-60% weakly			-25% weakly		-10% weakly		
IOP16c					5% weakly		-65% strongly	-25% strongly
IOP18	100% strongly		30% strongly	-5% strongly	5% strongly		10% strongly	25% strongly
IOP19						15% strongly		-10% strongly

Figure A.1: Table of HyMeX IOPs affecting the eight target areas. The upper box for each event shows the relative precipitation variation $(AS - CTRL)/CTRL$ in %, for the 7km grid. The lower box shows the classification of the extent of synoptic forcing by means of the Convective Adjustment time-scale (τ) as represented by the CTRL run. Typically similar τ take place for the same events in the CTRL and AS runs

2.8km	CA	VA	BA	CV	CO	LT	CI	NEI
IOP1				50% strongly				
IOP2								25% weakly
IOP4							50% strongly	
IOP6				-10% weakly		-75% weakly		
IOP7a				110% strongly				
IOP7b						-10% weakly		
IOP8		-7% strongly	10% weakly	-50% strongly				
IOP12a				-65% strongly	-15% strongly	33% strongly	25% weakly	
IOP12b			-65% weakly					
IOP13				-45% strongly	40% strongly	-40% strongly	15% weakly	0% strongly
IOP14				-10% strongly				
IOP15a	-100% weakly	0% strongly	-10% weakly					
IOP15b				-20% strongly				
IOP15c					140% weakly			
IOP16a	-5% weakly			-50% strongly		12% strongly		
IOP16c					50% weakly		-40% weakly	-35% strongly
IOP18	25% strongly		300% strongly	25% strongly	10% strongly		15% strongly	50% strongly
IOP19						-5% strongly		-20% strongly

Figure A.2: As Tab. A.1 but for the 2.8 km resolution

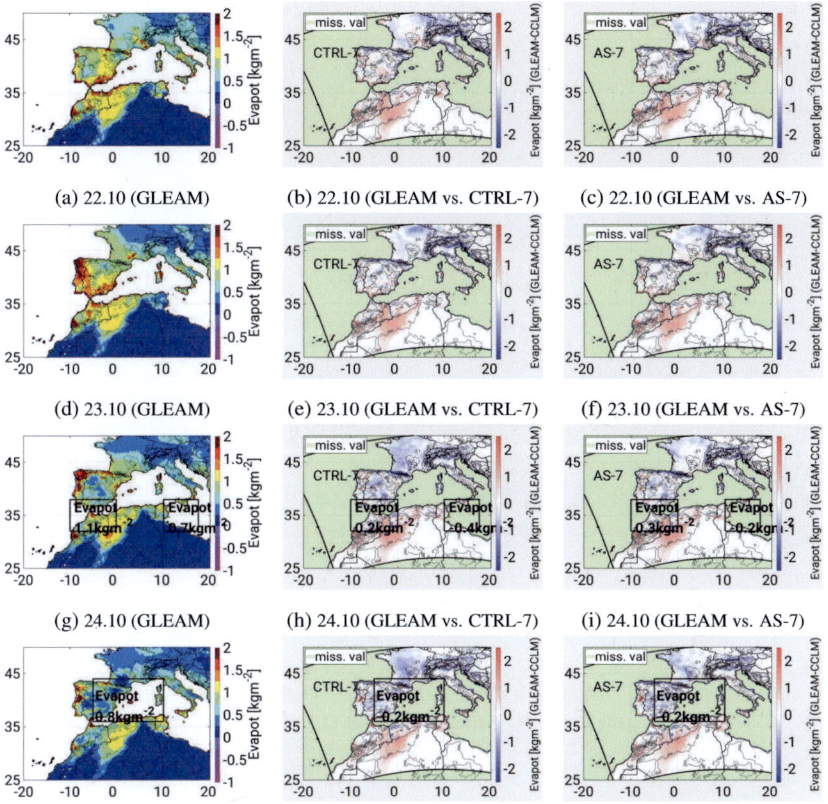

(a) 22.10 (GLEAM) (b) 22.10 (GLEAM vs. CTRL-7) (c) 22.10 (GLEAM vs. AS-7)

(d) 23.10 (GLEAM) (e) 23.10 (GLEAM vs. CTRL-7) (f) 23.10 (GLEAM vs. AS-7)

(g) 24.10 (GLEAM) (h) 24.10 (GLEAM vs. CTRL-7) (i) 24.10 (GLEAM vs. AS-7)

(j) 25.10 (GLEAM) (k) 25.10 (GLEAM vs. CTRL-7) (l) 25.10 (GLEAM vs. AS-7)

Figure A.3: Spatial distribution of the 0.25°x0.25° daily evapotranspiration GLEAM product (left) and differences against CTRL-7 and AS-7. Both simulation grids have been upscaled to the GLEAM grid size. CTRL-2.8 and AS-2.8 are not shown given they do not cover northern Africa. The shown dates, from 21-Oct to 25-Oct, show the intensive evapotranspiration over the Atlas mountains and to the south.

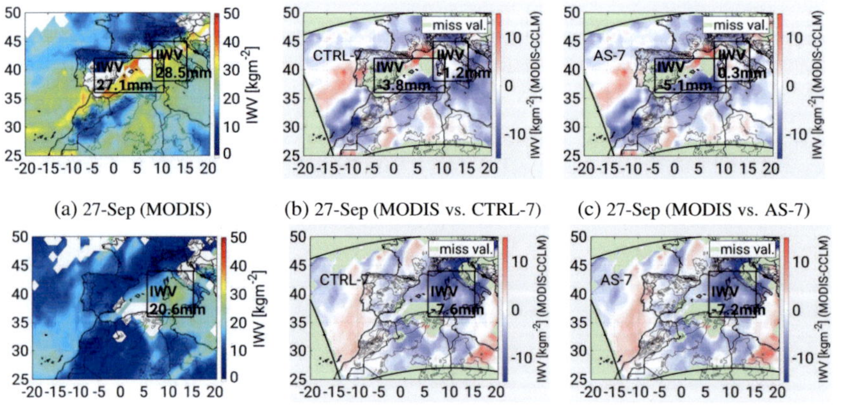

(a) 27-Sep (MODIS) (b) 27-Sep (MODIS vs. CTRL-7) (c) 27-Sep (MODIS vs. AS-7)

(d) 12-Nov (MODIS) (e) 12-Nov (MODIS vs. CTRL-7) (f) 12-Nov (MODIS vs. AS-7)

Figure A.4: Spatial distribution of daily averaged IWV as measured by the Moderate Resolution Imaging Spectroradiometer (MODIS) instrument aboard the Terra satellite (top) and difference plots of the same quantity against CTRL-7 and AS-7 (bottom) on the 27-Sep-2012 and 12-Nov-2012. The dates are representative of the large IWV transport originating at the northern Iberian Peninsula and over north Africa, respectively. Both COSMO-CLM simulations have been upscaled to the MODIS grid of 1°x1°.

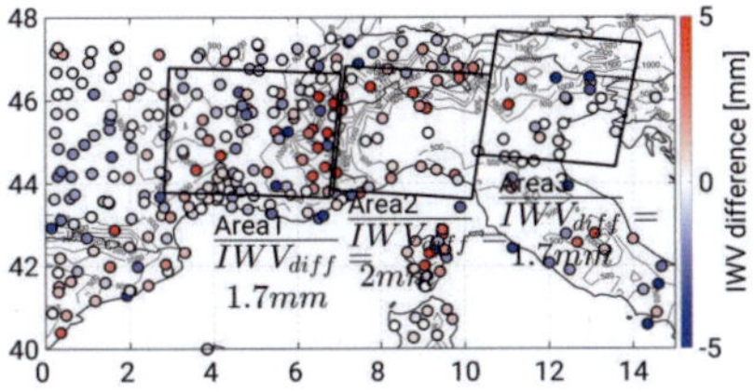

(a) AS-GPS-7

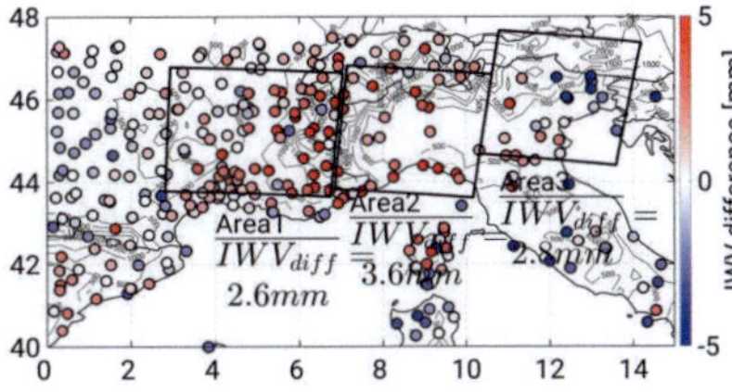

(b) AS-RAD-7

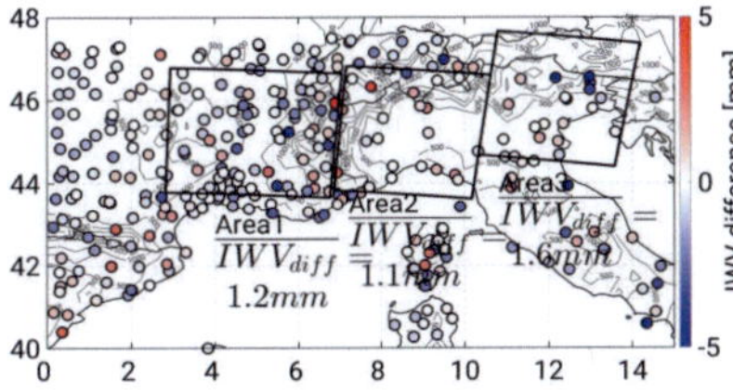

(c) AS-GPS-2.8

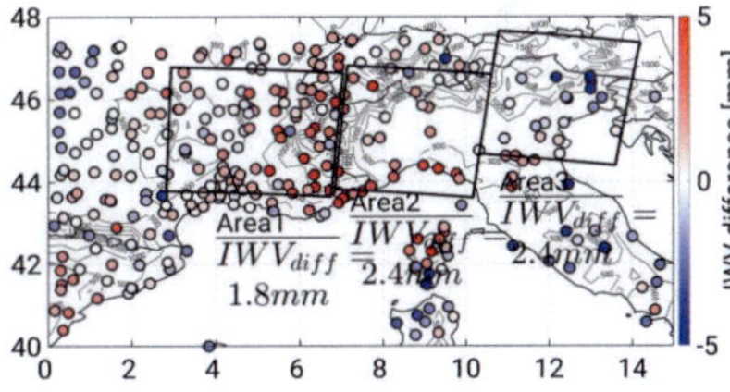

(d) AS-RAD-2.8

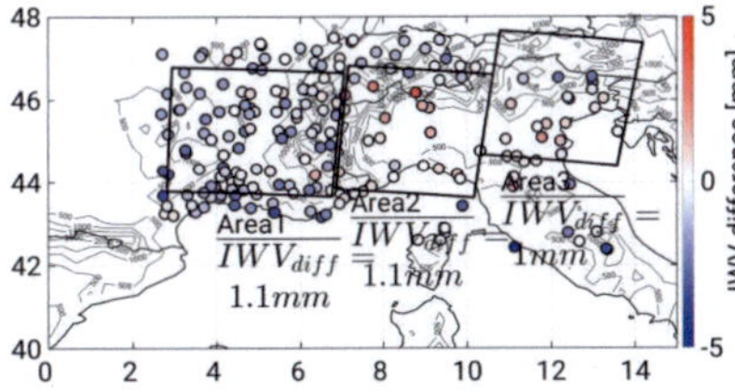

(e) AS-GPS-500

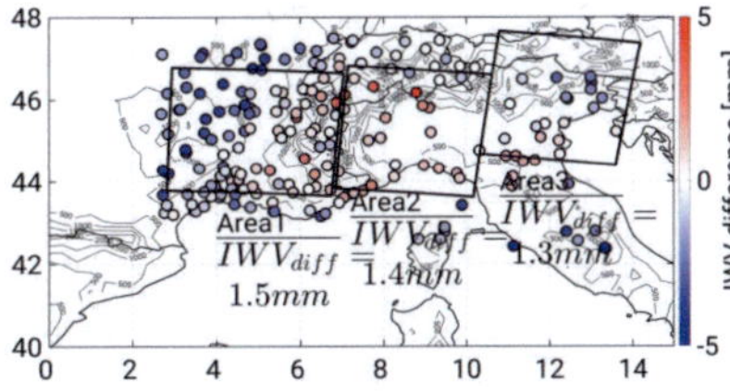

(f) AS-RAD-500

Figure A.5: Spatial distribution of the differences in IWV between GPS and COSMO for AS-GPS-7 (a), AS-RAD-7 (b), AS-GPS-2.8 (c), AS-RAD-2.8 (d), AS-GPS-500 (e) and AS-RAD-500 (f). The differences are obtained as (OBS-COSMO) after averaging IWV for every GPS station during the preconditioning period i.e. 23-Sep 12:00h to 23-Sep 19:00h. The IWV given by COSMO is obtained by natural interpolation from the neighbouring grid points.

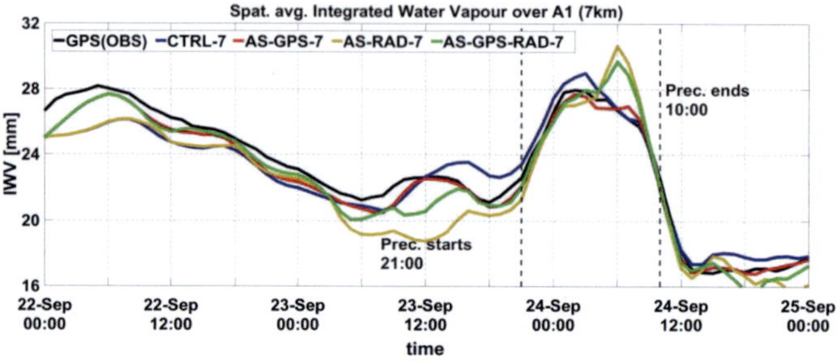

(a) Area1 (7km)

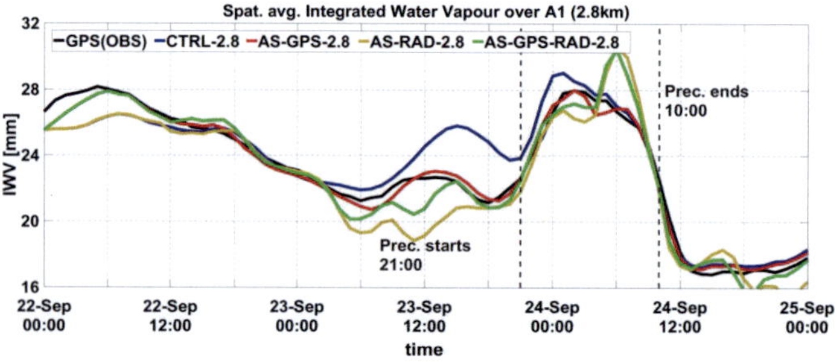

(b) Area1 (2.8 km)

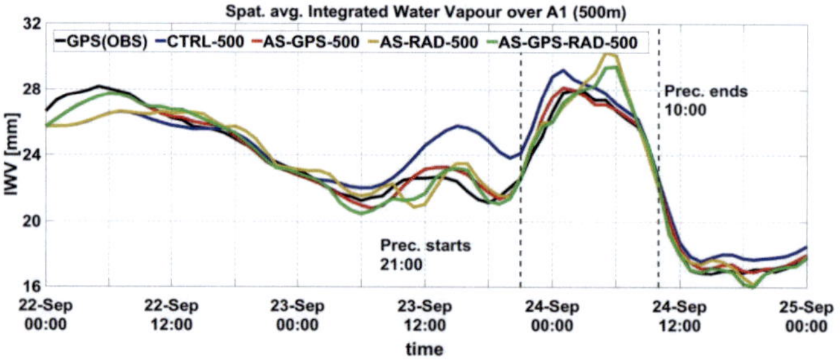

(c) Area1 (500m)

Figure A.6: Temporal evolution of spatially averaged IWV over investigation Area1 (see Fig. 5.4) and the simulations of HyMeX IOP6 (see Tab. 5.3). All GPS observations of Integrated Water vapour (IWV) within the investigation area are averaged and represented in black. The colours stand for the different COSMO simulations. The IWV values at the model grid points closest to the GPS stations are interpolated to the location of the GPS stations.

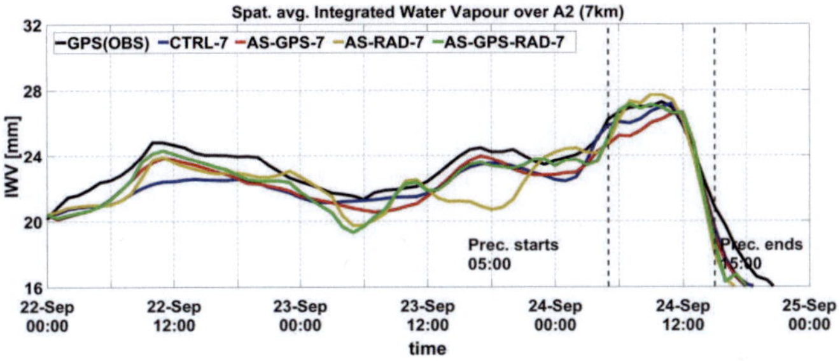

(a) Area2 (7km)

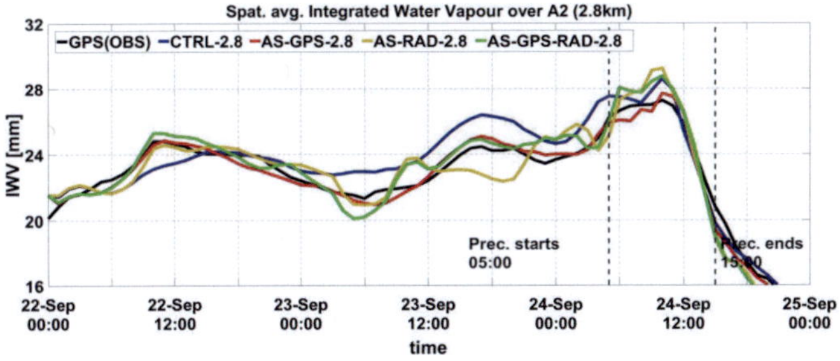

(b) Area2 (2.8km)

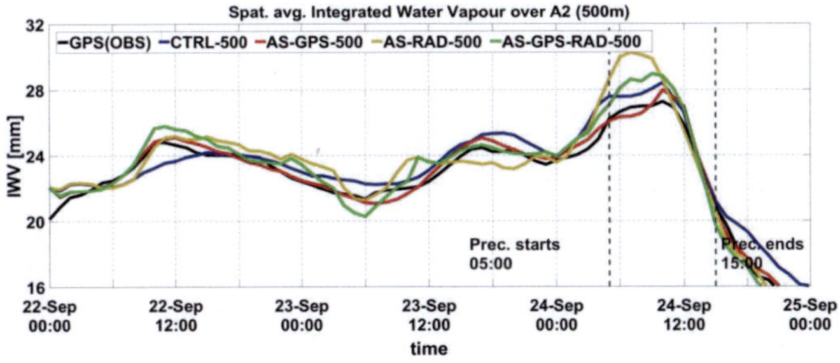

(c) Area2 (500m)

Figure A.7: As Figure A.6 but for investigation Area2

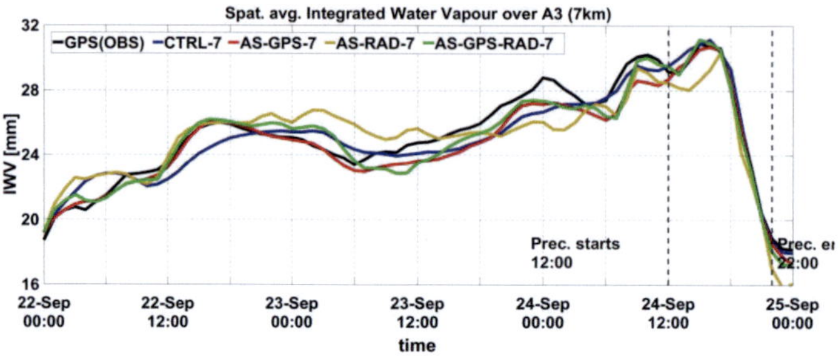

(a) Area3 (7km)

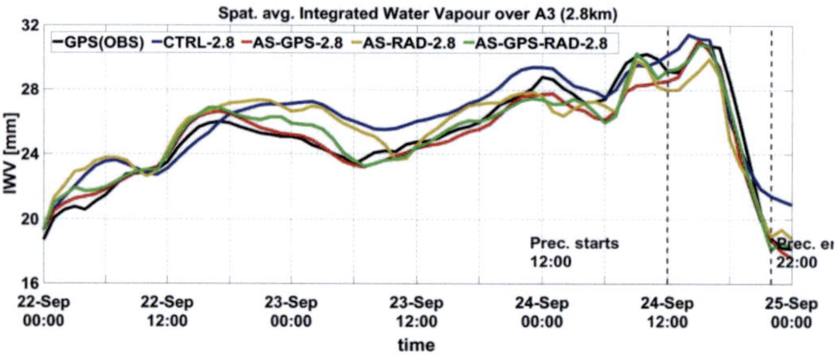

(b) Area3 (2.8km)

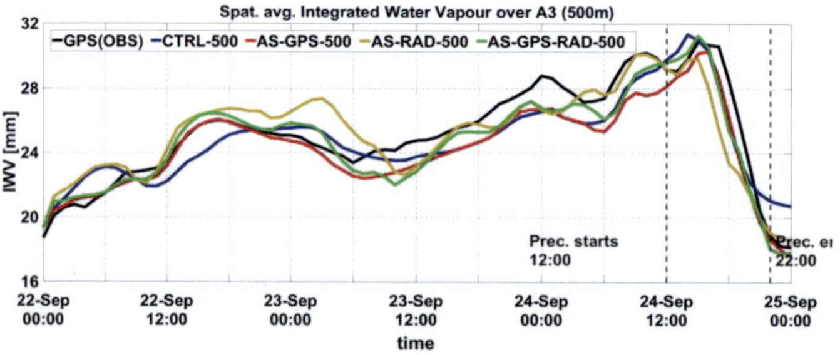

(c) Area3 (500m)

Figure A.8: As Figure A.6 but for investigation Area3

Table A.1: Structure Location Amount (SAL) metrics of the COSMO vs. CMORPH comparison. Daily accumulated precipitation are compared. See figure 8.4a for COSMO and 8.3 for CMORPH precipitation fields. The rows show the different observations nudged (AS), or no observations (CTRL).

Area1	**7km**			**2.8km**			**500m**		
	S	A	L	S	A	L	S	A	L
CTRL	-0.28	-0.64	0.08	-0.47	-0.59	0.1	-0.63	-0.63	0.08
AS-GPS	-0.16	-0.85	0.07	-0.71	-0.69	0.06	-0.6	-0.8	0.06
AS-RAD	-0.16	-0.27	0.1	-0.4	-0.24	0.08	-0.45	-0.4	0.12
AS-GPS-RAD	-0.33	-0.57	0.12	-0.48	-0.34	0.09	-0.58	-0.57	0.1
Area2	**7km**			**2.8km**			**500m**		
	S	A	L	S	A	L	S	A	L
CTRL	-0.05	-0.41	0.19	-0.17	-0.61	0.2	-0.51	-0.68	0.21
AS-GPS	-0.74	-0.76	0.3	-0.09	-0.77	0.17	-0.39	-0.92	0.16
AS-RAD	-0.23	-0.06	0.17	0.32	-0.14	0.22	-0.1	-0.31	0.13
AS-GPS-RAD	-0.36	-0.54	0.21	-0.06	-0.4	0.19	-0.2	-0.5	0.13
Area3	**7km**			**2.8km**			**500m**		
	S	A	L	S	A	L	S	A	L
CTRL	-0.64	-0.48	0.14	-0.56	-0.61	0.13	-0.24	-0.83	0.09
AS-GPS	-0.93	-0.9	0.13	-0.15	-0.73	0.11	-0.31	-1.16	0.15
AS-RAD	0.09	-0.73	0.18	-0.76	-1.02	0.16	-1.09	-0.84	0.12
AS-GPS-RAD	-0.16	-0.83	0.14	-0.39	-0.9	0.13	-0.99	-0.85	0.11

Acronyms

T_d Dew Point Temperature.

τ Convective Adjustment time-scale.

θ_e Equivalent Potential Temperature.

q_v Specific Humidity.

AI Agreement Index.

AIRS Atmospheric Infrared Sounder.

AMSU Advanced Microwave Sounding Unit.

ARPS Advanced Regional Prediction System.

BA Balearic Islands.

BLPBs Boundary Layer Pressurized Balloons.

BOLAM Bologna Limited-Area Model.

CA Catalonia.

CAPE Convective Available Potential Energy.

CDF Cumulative Distribution Function.

CI Central Italy.

CIN Convective Inhibition.

CMORPH Climate Prediction Center Morphing Technique.

CMPs Convection Permitting Models.

CO Corsica.

CORDEX Coordinated Regional Climate Downscaling Experiment.

COSMO Consortium for Small-scale Modelling.

COSMO-CLM COSMO in CLimate Mode.

CV Cevennes-Vivarais.

D Detrainment.

DA Data Assimilation.

DMC Deep Moist Convection.

E Entrainment.

ECMWF European Centre for Medium-range Weather Forecasts.

EL Equilibrium Level.

ENSO El Niño Southern Oscillation.

FR France.

GLEAM Global Land Evaporation Amsterdam Model.

GPS Global Positioning System.

HIRLAM High Resolution Limited Area Model.

HPE Heavy Precipitation Event.

HyMeX HYdrological cycle in the Mediterranean eXperiment.

IFS Integrated Forecast System.

IOP Intensive Observation Period.

IP Iberian Peninsula.

IT Italy.

ITCZ Intertropical Convergence Zone.

IWV Integrated Water vapour.

KENDA Kalman filter for convective-scale Data Assimilation.

LCL Lifted Condensation Level.

LFC Level of Free Convection.

LT Liguria-Tuscany.

MAE Mean Absolute Error.

MCSs Mesoscale Convective Systems.

MM5 Fifth-Generation Penn State/BCAR Mesoscale Model.

MODIS Moderate Resolution Imaging Spectroradiometer.

MOLOCH MOdello LOCale.

NAO North Atlantic Oscillation.

NEI North-eastern Italy.

NWP Numerical Weather Prediction.

PBL Planetary Boundary Layer.

PDF Probability Distribution Function.

PMSL Pressure at the Mean Sea Level.

R Correlation Coefficient.

RCM Regional Climate Model.

RMSE Root Mean Square Error.

SAL Structure Location Amount.

SCAND Scandinavian Pattern.

SOP1 Special Observation Period 1.

SST Sea Surface Temperature.

TERRA-ML Multi-Layer Soil and Vegetation Model.

VA Valencia.

WMed Western Mediterranean.

ZTD Zenith Total Delay.

Bibliography

Ahrens, C. D., and P. J. Samson, 2010: *Extreme Weather and Climate*. Cengage Learning.

Alfieri, L., F. Dottori, R. Betts, P. Salamon, and L. Feyen, 2018: Multi-model projections of river flood risk in europe under global warming. *Climate*, **6 (1)**, 6, doi:10.3390/cli6010006.

Allan, R. P., and B. J. Soden, 2008: Atmospheric warming and the amplification of precipitation extremes. *Science*, **321 (5895)**, 1481–1484, doi: 10.1126/science.1160787.

Amengual, A., V. Homar, and O. Jaume, 2015: Potential of a probabilistic hydrometeorological forecasting approach for the 28 september 2012 extreme flash flood in murcia, spain. *Atmospheric Research*, **166**, 10–23, doi: 10.1016/j.atmosres.2015.06.012.

Andersson, T., M. Andersson, C. Jacobsson, and S. Nilsson, 1989: Thermodynamic indices for forecasting thunderstorms in southern sweden. *Meteorological Magazine*, **118 (1404)**.

Argüeso, D., J. P. Evans, L. Fita, and K. J. Bormann, 2013: Temperature response to future urbanization and climate change. *Climate Dynamics*, **42 (7-8)**, 2183–2199, doi:10.1007/s00382-013-1789-6.

Aumann, H. H., and Coauthors, 2003: Airs/amsu/hsb on the aqua mission: design, science objectives, data products, and processing systems. *IEEE Transactions on Geoscience and Remote Sensing*, **41 (2)**, 253–264, doi: 10.1109/TGRS.2002.808356.

Barthlott, C., and C. Hoose, 2015: Spatial and temporal variability of clouds and precipitation over germany: multiscale simulations across the &quotgray zone&quot. *Atmospheric Chemistry and Physics*, **15 (21)**, 12 361–12 384, doi:10.5194/acp-15-12361-2015.

Barthlott, C., and N. Kalthoff, 2011: A numerical sensitivity study on the impact of soil moisture on convection-related parameters and convective precipitation over complex terrain. *Journal of the Atmospheric Sciences*, **68 (12)**, 2971–2987, doi:10.1175/jas-d-11-027.1.

Bevis, M., S. B. T. A. Herring, C. Rocken, R. A. Anthes, and R. H. Ware, 1992: Gps meteorology: Remote sensing of atmospheric water vapor using the global positioning system. *Journal of Geophysical Research*, **97 (D14)**, 15.787–15.801.

Birch, C. E., D. J. Parker, J. H. Marsham, D. Copsey, and L. Garcia-Carreras, 2014: A seamless assessment of the role of convection in the water cycle of the west african monsoon. *Journal of Geophysical Research: Atmospheres*, **119 (6)**, 2890–2912, doi:10.1002/2013jd020887.

Bock, O., P. Bosser, R. Pacione, M. Nuret, N. Fourrié, and A. Parracho, 2016: A high-quality reprocessed ground-based GPS dataset for atmospheric process studies, radiosonde and model evaluation, and reanalysis of HyMeX special observing period. *Quarterly Journal of the Royal Meteorological Society*, **142**, 56–71, doi:10.1002/qj.2701.

Bock, O., M.-N. Bouin, A. Walpersdorf, J. P. Lafore, S. Janicot, F. Guichard, and A. Agusti-Panareda, 2007: Comparison of ground-based GPS precipitable water vapour to independent observations and NWP model reanalyses over africa. *Quarterly Journal of the Royal Meteorological Society*, **133 (629)**, 2011–2027, doi:10.1002/qj.185.

Boehm, J., B. Werl, and H. Schuh, 2006: Troposphere mapping functions for GPS and very long baseline interferometry from european centre for medium-range weather forecasts operational analysis data. *Journal of Geophysical Research: Solid Earth*, **111 (B2)**, n/a–n/a, doi:10.1029/2005jb003629.

Bolle, H.-J., Ed., 2003: *Mediterranean Climate*. Springer Berlin Heidelberg, doi:10.1007/978-3-642-55657-9.

Bouin, M.-N., J.-L. Redelsperger, and C. L. Brossier, 2017: Processes leading to deep convection and sensitivity to sea-state representation during HyMeX IOP8 heavy precipitation event. *Quarterly Journal of the Royal Meteorological Society*, **143 (707)**, 2600–2615, doi:10.1002/qj.3111.

Caldas-Álvarez, A., S. Khodayar, and O. Bock, 2017: GPS – zenith total delay assimilation in different resolution simulations of a heavy precipitation event over southern france. *Advances in Science and Research*, **14**, 157–162, doi:10.5194/asr-14-157-2017.

Chan, S. C., E. J. Kendon, H. J. Fowler, S. Blenkinsop, C. A. T. Ferro, and D. B. Stephenson, 2012: Does increasing the spatial resolution of a regional climate model improve the simulated daily precipitation? *Climate Dynamics*, **41 (5-6)**, 1475–1495, doi:10.1007/s00382-012-1568-9.

Chazette, P., C. Flamant, J.-C. Raut, J. Totems, and X. Shang, 2015a: Tropical moisture enriched storm tracks over the mediterranean and their link with intense rainfall in the cevennes-vivarais area during HyMeX. *Quarterly Journal of the Royal Meteorological Society*, **142**, 320–334, doi:10.1002/qj.2674.

Chazette, P., and Coauthors, 2015b: A multi-instrument and multi-model assessment of atmospheric moisture variability over the western mediterranean during HyMeX. *Quarterly Journal of the Royal Meteorological Society*, **142**, 7–22, doi:10.1002/qj.2671.

Coppola, E., and Coauthors, 2018: A first-of-its-kind multi-model convection permitting ensemble for investigating convective phenomena over europe and the mediterranean. *Climate Dynamics*, doi:10.1007/s00382-018-4521-8.

Cress, A., H. Anlauf, H. W. Bitzer, A. Rhodin, C. Schraff, K. Helmert, and K. Stephan, 2012: Global and regional impact studies at the german weather service (dwd). *5th WMO Workshop on the Impact of Various observing systems on NWP*, Sedona (USA).

Crook, N. A., 1996: Sensitivity of moist convection forced by boundary layer processes to low-level thermodynamic fields. *Monthly Weather Review*, **124 (8)**, 1767–1785, doi:10.1175/1520-0493(1996)124<1767:somcfb>2.0.co;2.

Dacre, H. F., P. A. Clark, O. Martinez-Alvarado, M. A. Stringer, and D. A. Lavers, 2015: How do atmospheric rivers form? *Bulletin of the American Meteorological Society*, **96 (8)**, 1243–1255, doi:10.1175/bams-d-14-00031.1.

Dafis, S., K. Lagouvardos, V. Kotroni, T. M. Giannaros, and A. Bartzokas, 2017: Observational and modeling study of a mesoscale convective system during the HyMeX — SOP1. *Atmospheric Research*, **187**, 1–15, doi:10.1016/j.atmosres.2016.12.001.

Davolio, S., A. Volonté, A. Manzato, A. Pucillo, A. Cicogna, and M. E. Ferrario, 2016: Mechanisms producing different precipitation patterns over north-eastern italy: insights from HyMeX-SOP1 and previous events. *Quarterly Journal of the Royal Meteorological Society*, **142**, 188–205, doi:10.1002/qj.2731.

Defer, E., and Coauthors, 2014: Properties of the lightning activity at storm scale during hymex sop1 campaign: Comparison between an isolated storm (05 sept 2012), a multi-cellular system (24 sept 2012) and a tornadic cell (14 oct 2012). *XV International Conference on Atmospheric Electricity*.

Devidasrao, S., 2012: Diagnostic verification of atmospheric water cycle predicted by regional mesoscale models and ensemble systems. Phd thesis, Universität Hamburg.

Doms, G., and Coauthors, 2011: *A Description of the Nonhydrostatic Regional COSMO Model Part II : Physical Parameterization*. Deutsche Wetterdienst.

Doswell, C. A., H. E. Brooks, and R. A. Maddox, 1996: Flash flood forecasting: An ingredients-based methodology. *Weather and Forecasting*, **11 (4)**, 560–581, doi:10.1175/1520-0434(1996)011<0560:fffaib>2.0.co;2.

Drobinski, P., V. Ducroq, and K. Beranger, 2007: Hymex white book.

Drobinski, P., and Coauthors, 2016: Scaling precipitation extremes with temperature in the mediterranean: past climate assessment and projection in anthropogenic scenarios. *Climate Dynamics*, **51 (3)**, 1237–1257, doi:10.1007/s00382-016-3083-x.

Ducrocq, V., S. Davolio, R. Ferretti, C. Flamant, V. H. Santaner, N. Kalthoff, E. Richard, and H. Wernli, 2016: Introduction to the HyMeX special issue on 'advances in understanding and forecasting of heavy precipitation in the mediterranean through the HyMeX SOP1 field campaign'. *Quarterly Journal of the Royal Meteorological Society*, **142**, 1–6, doi:10.1002/qj.2856.

Ducrocq, V., and Coauthors, 2014: HyMeX-SOP1: The field campaign dedicated to heavy precipitation and flash flooding in the northwestern mediterranean. *Bulletin of the American Meteorological Society*, **95 (7)**, 1083–1100, doi:10.1175/bams-d-12-00244.1.

Duffourg, F., and V. Ducrocq, 2011: Origin of the moisture feeding the heavy precipitating systems over southeastern france. *Natural Hazards and Earth System Science*, **11 (4)**, 1163–1178, doi:10.5194/nhess-11-1163-2011.

Duffourg, F., and Coauthors, 2016: Offshore deep convection initiation and maintenance during the HyMeX IOP 16a heavy precipitation event. *Quar-*

terly Journal of the Royal Meteorological Society, **142**, 259–274, doi: 10.1002/qj.2725.

Duruisseau, F., and Coauthors, 2017: Investigating the potential benefit to a mesoscale NWP model of a microwave sounder on board a geostationary satellite. *Quarterly Journal of the Royal Meteorological Society*, **143 (706)**, 2104–2115, doi:10.1002/qj.3070.

EEA, 2016: Climate change, impacts and vulnerability in europe 2016. an indicator-based report. Indicator-based report, European Environment Agency, European Environment Agency, Kongens Nytorv 6 1050 Copenhagen K Denmark.

Emanuel, K. A., 1994: *Atmospheric Convection*. Oxford University Press.

FAO, 1995: *The Digitized Soil Map of the World Including Derived Soil Properties (version 3.6)*. Rome, Italy, FAO.

Ferretti, R., and Coauthors, 2014: Overview of the first HyMeX special observation period over italy: observations and model results. *Hydrology and Earth System Sciences*, **18 (5)**, 1953–1977, doi:10.5194/hess-18-1953-2014.

Fischer, E. M., and R. Knutti, 2016: Observed heavy precipitation increase confirms theory and early models. *Nature Climate Change*, **6 (11)**, 986–991, doi:10.1038/nclimate3110.

Fosser, G., S. Khodayar, and P. Berg, 2014: Benefit of convection permitting climate model simulations in the representation of convective precipitation. *Climate Dynamics*, **44 (1-2)**, 45–60, doi:10.1007/s00382-014-2242-1.

Fosser, G., S. Khodayar, and P. Berg, 2016: Climate change in the next 30 years: What can a convection-permitting model tell us that we did not already know? *Climate Dynamics*, **48 (5-6)**, 1987–2003, doi:10.1007/s00382-016-3186-4.

Gantner, L., V. Maurer, N. Kalthoff, and O. Kiseleva, 2017: The impact of land-surface parameter properties and resolution on the simulated cloud-topped atmospheric boundary layer. *Boundary-Layer Meteorology*, **165 (3)**, 475–496, doi:10.1007/s10546-017-0286-6.

Gao, B.-C., and Y. J. Kaufman, 2003: Water vapor retrievals using moderate resolution imaging spectroradiometer (MODIS) near-infrared channels. *Journal of Geophysical Research: Atmospheres*, **108 (D13)**, n/a–n/a, doi: 10.1029/2002jd003023.

Gaume, E., M. Borga, M. C. Llassat, S. Maouche, M. Lang, and M. Diakakis, 2016: Mediterranean extreme floods and flash floods. *IRD Editions*, **The Mediterranean Region under Climate Change. A scientific update**, 133–144.

Girolamo, P. D., and Coauthors, 2017: Lidar observations of low-level wind reversals over the gulf of lion and characterization of their impact on the water vapour variability. Author(s), doi:10.1063/1.4975532.

González-Rojí, S. J., J. Sáenz, G. Ibarra-Berastegi, and J. D. de Argandoña, 2018: Moisture balance over the iberian peninsula according to a regional climate model: The impact of 3dvar data assimilation. *Journal of Geophysical Research: Atmospheres*, **123 (2)**, 708–729, doi:10.1002/2017jd027511.

Greve, P., and S. I. Seneviratne, 2015: Assessment of future changes in water availability and aridity. *Geophysical Research Letters*, **42 (13)**, 5493–5499, doi:10.1002/2015gl064127.

Guerova, G., and Coauthors, 2016: Review of the state of the art and future prospects of the ground-based GNSS meteorology in europe. *Atmospheric Measurement Techniques*, **9 (11)**, 5385–5406, doi:10.5194/amt-9-5385-2016.

Hally, A., E. Richard, and V. Ducrocq, 2014: An ensemble study of HyMeX IOP6 and IOP7a: sensitivity to physical and initial and boundary condition uncertainties. *Natural Hazards and Earth System Sciences*, **14 (5)**, 1071–1084, doi:10.5194/nhess-14-1071-2014.

Hastings, D. A., and Coauthors, 1999: *The Global Land One-kilometer Base Elevation (GLOBE) Digital Elevation Model, Version 1.0*. National Geophysical Data Center, 325 Broadway, Boulder, Colorado 80305-3328, U.S.A., National Oceanic and Atmospheric Administration.

Hdidou, F., S. Mordane, and S. Sbii, 2016: Installation and processing of ground-based gps stations of the moroccan national meteorological service. Technical report, Direction de la Météorologie Nationale CNRMSI.

Hohenegger, C., P. Brockhaus, C. S. Bretherton, and C. Schär, 2009: The soil moisture–precipitation feedback in simulations with explicit and parameterized convection. *Journal of Climate*, **22 (19)**, 5003–5020, doi:10.1175/2009jcli2604.1.

Hohenegger, C., and B. Stevens, 2013: Preconditioning deep convection with cumulus congestus. *Journal of the Atmospheric Sciences*, **70 (2)**, 448–464, doi:10.1175/jas-d-12-089.1.

Holton, J. R., and G. J. Hakim, 2012: *An Introduction to Dynamic Meteorology*. Elsevier LTD, Oxford.

Honda, T., and T. Kawano, 2015: How does mid-tropospheric dry air affect the evolution of supercellular convection? *Atmospheric Research*, **157**, 1–16, doi:10.1016/j.atmosres.2015.01.015.

Houze, R. A., 2012: Orographic effects on precipitating clouds. *Reviews of Geophysics*, **50 (1)**, doi:10.1029/2011rg000365.

Huang, Q., J. H. Marsham, D. J. Parker, W. Tian, and T. Weckwerth, 2009: A comparison of roll and nonroll convection and the subsequent deepening moist convection: An LEM case study based on SCMS data. *Monthly Weather Review*, **137 (1)**, 350–365, doi:10.1175/2008mwr2450.1.

Huffman, G. J., and Coauthors, 2007: The trmm multisatellite precipitation analysis (tmpa): Quasi-global, multiyear, combined-sensor precipitation estimates at fine scales. *Journal of Hydrometeorology*, **8 (1)**, 38–55, doi: 10.1175/JHM560.1.

Hunt, B. R., E. J. Kostelich, and I. Szunyogh, 2007: Efficient data assimilation for spatiotemporal chaos: A local ensemble transform kalman filter. *Physica D: Nonlinear Phenomena*, **230 (1-2)**, 112–126, doi:10.1016/j.physd.2006.11.008.

IPCC, 2018: Global warming of 1.5°c. Special report, Intergovernmental Panel on Climate Change.

Jacquette, E., A. A. Bitar, A. Mialon, Y. Kerr, A. Quesney, F. Cabot, and P. Richaume, 2010: SMOS CATDS level 3 global products over land. *Remote Sensing for Agriculture, Ecosystems, and Hydrology XII*, C. M. U. Neale, and A. Maltese, Eds., SPIE, doi:10.1117/12.865093.

Joyce, R. J., J. E. Janowiak, P. A. Arkin, and P. Xie, 2004: CMORPH: A method that produces global precipitation estimates from passive microwave and infrared data at high spatial and temporal resolution. *Journal of Hydrometeorology*, **5 (3)**, 487–503, doi:10.1175/1525-7541(2004)005<0487:camtpg>2.0.co;2.

Kain, J. S., and Coauthors, 2008: Some practical considerations regarding horizontal resolution in the first generation of operational convection-allowing NWP. *Weather and Forecasting*, **23 (5)**, 931–952, doi:10.1175/waf2007106.1.

Keil, C., F. Heinlein, and G. C. Craig, 2013: The convective adjustment time-scale as indicator of predictability of convective precipitation. *Quarterly Journal of the Royal Meteorological Society*, **140 (679)**, 480–490, doi: 10.1002/qj.2143.

Keil, C., A. Röpnack, G. C. Craig, and U. Schumann, 2008: Sensitivity of quantitative precipitation forecast to height dependent changes in humidity. *Geophysical Research Letters*, **35 (9)**, doi:10.1029/2008gl033657.

Kendon, E. J., N. M. Roberts, C. A. Senior, and M. J. Roberts, 2012: Realism of rainfall in a very high-resolution regional climate model. *Journal of Climate*, **25 (17)**, 5791–5806, doi:10.1175/jcli-d-11-00562.1.

Kendon, E. J., and Coauthors, 2017: Do convection-permitting regional climate models improve projections of future precipitation change? *Bulletin of the American Meteorological Society*, **98 (1)**, 79–93, doi:10.1175/bams-d-15-0004.1.

Khodayar, S., 2009: High-resolution analysis of the initiation of deep convection forced by boundary-layer processes. Phd thesis, Fakultät für Physik der Universität Karlsruhe.

Khodayar, S., B. Czajka, A. Caldas-Alvarez, S. Helgert, C. Flamant, P. D. Girolamo, O. Bock, and P. Chazette, 2018: Multi-scale observations of atmospheric moisture variability in relation to heavy precipitating systems in the north-western mediterranean during HyMeX IOP12. *Quarterly Journal of the Royal Meteorological Society*, doi:10.1002/qj.3402.

Khodayar, S., N. Kalthoff, and C. Kottmeier, 2016a: Atmospheric conditions associated with heavy precipitation events in comparison to seasonal means in the western mediterranean region. *Climate Dynamics*, **51 (3)**, 951–967, doi:10.1007/s00382-016-3058-y.

Khodayar, S., F. Raff, N. Kalthoff, and O. Bock, 2015: Diagnostic study of a high-precipitation event in the western mediterranean: adequacy of current operational networks. *Quarterly Journal of the Royal Meteorological Society*, **142**, 72–85, doi:10.1002/qj.2600.

Khodayar, S., and Coauthors, 2016b: A seamless weather-climate multi-model intercomparison on the representation of a high impact weather event in the western mediterranean: HyMeX IOP12. *Quarterly Journal of the Royal Meteorological Society*, **142**, 433–452, doi:10.1002/qj.2700.

Krichak, S. O., J. Barkan, J. S. Breitgand, S. Gualdi, and S. B. Feldstein, 2014: The role of the export of tropical moisture into midlatitudes for extreme precipitation events in the mediterranean region. *Theoretical and Applied Climatology*, **121 (3-4)**, 499–515, doi:10.1007/s00704-014-1244-6.

Köppen, W., 1900: Versuch einer klassifikation der klimate, vorzugsweise nach beziehungen zur pflanzenwelt. *Geographische Zeitschrift*.

Lamb, P. J., D. H. Portis, and A. Zangvil, 2012: Investigation of large-scale atmospheric moisture budget and land surface interactions over u.s. southern great plains including for CLASIC (june 2007). *Journal of Hydrometeorology*, **13 (6)**, 1719–1738, doi:10.1175/jhm-d-12-01.1.

Latifovic, R., Z.-L. Zhu, J. Cihlar, C. Giri, and I. Olthof, 2004: Land cover mapping of north and central america—global land cover 2000. *Remote Sensing of Environment*, **89 (1)**, 116–127, doi:10.1016/j.rse.2003.11.002.

Lee, K.-O., C. Flamant, V. Ducrocq, F. Duffourg, N. Fourrié, and S. Davolio, 2016: Convective initiation and maintenance processes of two back-building mesoscale convective systems leading to heavy precipitation events in southern italy during HyMeX IOP 13. *Quarterly Journal of the Royal Meteorological Society*, **142 (700)**, 2623–2635, doi:10.1002/qj.2851.

Lee, K.-O., C. Flamant, V. Ducrocq, F. Duffourg, N. Fourrié, J. Delanoë, and J. Bech, 2017: Initiation and development of a mesoscale convective system in the ebro river valley and related heavy precipitation over northeastern spain during HyMeX IOP 15a. *Quarterly Journal of the Royal Meteorological Society*, **143 (703)**, 942–956, doi:10.1002/qj.2978.

Lee, K.-O., C. Flamant, F. Duffourg, V. Ducrocq, and J.-P. Chaboureau, 2018: Impact of upstream moisture structure on a back-building convective precipitation system in south-eastern france during HyMeX IOP13. *Atmospheric Chemistry and Physics*, **18 (23)**, 16 845–16 862, doi:10.5194/acp-18-16845-2018.

Leuenberger, D., M. Stoll, and A. Roches, 2010: *Description of some convective indices implemented in the COSMO model*. MeteoSwiss.

Li, X., and X. Shen, 2010: Sensitivity of cloud-resolving precipitation simulations to uncertainty of vertical structures of initial conditions. *Quarterly Journal of the Royal Meteorological Society*, **136 (646)**, 201–212, doi:10.1002/qj.540.

Lim, Y.-K., L. B. Stefanova, S. C. Chan, S. D. Schubert, and J. J. O'Brien, 2010: High-resolution subtropical summer precipitation derived from dynamical downscaling of the NCEP/DOE reanalysis: how much small-scale information is added by a regional model? *Climate Dynamics*, **37 (5-6)**, 1061–1080, doi:10.1007/s00382-010-0891-2.

Lionello, P., 2012: The climate of the mediterranean region : From the past to the future. Elsevier Science, Burlington, 2.7.1 Late-Holocene Climate Records from Eastern Mediterranean Lake Sediments. - Description based on print version record.

Lionello, P., and L. Scarascia, 2018: The relation between climate change in the mediterranean region and global warming. *Regional Environmental Change*, **18 (5)**, 1481–1493, doi:10.1007/s10113-018-1290-1.

Manzato, A., S. Davolio, M. M. Miglietta, A. Pucillo, and M. Setvák, 2015: 12 september 2012: A supercell outbreak in NE italy? *Atmospheric Research*, **153**, 98–118, doi:10.1016/j.atmosres.2014.07.019.

Markowski, P., and Y. Richardson, 2010: *Mesoscale Meteorology in Midlatitudes*. Wiley.

Martens, B., and Coauthors, 2017: Gleam v3: satellite-based land evaporation and root-zone soil moisture. *Geoscientific Model Development*, **10 (5)**, 1903–1925, doi:10.5194/gmd-10-1903-2017.

Martinet, M., O. Nuissier, F. Duffourg, V. Ducrocq, and D. Ricard, 2017: Fine-scale numerical analysis of the sensitivity of the HyMeX IOP16a heavy precipitating event to the turbulent mixing-length parametrization. *Quarterly Journal of the Royal Meteorological Society*, **143 (709)**, 3122–3135, doi: 10.1002/qj.3167.

Mohr, K. I., J. S. Famiglietti, and E. J. Zipser, 1999: The contribution to tropical rainfall with respect to convective system type, size, and intensity estimated from the 85-GHz ice-scattering signature. *Journal of Applied Meteorology*, **38 (5)**, 596–606, doi:10.1175/1520-0450(1999)038<0596:tcttrw>2.0.co;2.

Molini, L., A. Parodi, N. Rebora, and G. C. Craig, 2011: Classifying severe rainfall events over italy by hydrometeorological and dynamical criteria. *Quarterly Journal of the Royal Meteorological Society*, **137 (654)**, 148–154, doi:10.1002/qj.741.

Moncrieff, M. W., and M. J. Miller, 1976: The dynamics and simulation of tropical cumulonimbus and squall lines. *Quarterly Journal of the Royal Meteorological Society*, **102 (432)**, 373–394, doi:10.1002/qj.49710243208.

Munn, T., M. C. MacXracken, and J. S. Perry, 2003: *Encyclopedia of Global Environmental Change, The Earth System: Physical and Chemical Dimensions of Global Environmental Change (Volume 1)*. Wiley.

Pastor, F., J. A. Valiente, and J. L. Palau, 2017: Sea surface temperature in the mediterranean: Trends and spatial patterns (1982–2016). *Pure and Applied Geophysics*, **175 (11)**, 4017–4029, doi:10.1007/s00024-017-1739-z.

Pfahl, S., P. A. O'Gorman, and E. M. Fischer, 2017: Understanding the regional pattern of projected future changes in extreme precipitation. *Nature Climate Change*, **7 (6)**, 423–427, doi:10.1038/nclimate3287.

Prein, A. F., and Coauthors, 2015: A review on regional convection-permitting climate modeling: Demonstrations, prospects, and challenges. *Reviews of Geophysics*, **53 (2)**, 323–361, doi:10.1002/2014rg000475.

Ramos, A. M., M. J. Martins, R. Tomé, and R. M. Trigo, 2018: Extreme precipitation events in summer in the iberian peninsula and its relationship with atmospheric rivers. *Frontiers in Earth Science*, **6**, doi:10.3389/feart.2018.00110.

Rasmusson, E. M., 1968: ATMOSPHERIC WATER VAPOR TRANSPORT AND THE WATER BALANCE OF NORTH AMERICA. *Monthly Weather Review*, **96 (10)**, 720–734, doi:10.1175/1520-0493(1968)096<0720:awvtat>2.0.co;2.

Raveh-Rubin, S., and H. Wernli, 2015: Large-scale wind and precipitation extremes in the mediterranean: a climatological analysis for 1979-2012. *Quarterly Journal of the Royal Meteorological Society*, **141 (691)**, 2404–2417, doi:10.1002/qj.2531.

Ricard, D., V. Ducrocq, and L. Auger, 2012: A climatology of the mesoscale environment associated with heavily precipitating events over a northwestern mediterranean area. *Journal of Applied Meteorology and Climatology*, **51 (3)**, 468–488, doi:10.1175/jamc-d-11-017.1.

Röhner, L., K.-U. Nerding, and U. Corsmeier, 2016: Diagnostic study of a HyMeX heavy precipitation event over spain by investigation of moisture trajectories. *Quarterly Journal of the Royal Meteorological Society*, **142**, 287–297, doi:10.1002/qj.2825.

Sala, J. Q., A. G. Olcina, A. P. Cuevas, J. O. Cantos, A. R. Amoros, and E. M. Chiva, 2000: Climatic warming in the spanish mediterranean: Natural trend or urban effect. *Climatic Change*, **46 (4)**, 473–483, doi:10.1023/a:1005688608044.

Scheffknecht, P., E. Richard, and D. Lambert, 2016: A highly localized high-precipitation event over corsica. *Quarterly Journal of the Royal Meteorological Society*, **142**, 206–221, doi:10.1002/qj.2795.

Schiro, K. A., J. D. Neelin, D. K. Adams, and B. R. Lintner, 2016: Deep convection and column water vapor over tropical land versus tropical ocean: A comparison between the amazon and the tropical western pacific. *Journal of the Atmospheric Sciences*, **73 (10)**, 4043–4063, doi:10.1175/jas-d-16-0119.1.

Schraff, C., and R. Hess, 2012: *A Description of the Nonhydrostatic Regional COSMO Model Part III : Data Assimilation*. Deutsche Wetterdienst.

Schraff, C., H. Reich, A. Rhodin, A. Schomburg, K. Stephan, A. Periáñez, and R. Potthast, 2016: Kilometre-scale ensemble data assimilation for the COSMO model (KENDA). *Quarterly Journal of the Royal Meteorological Society*, **142 (696)**, 1453–1472, doi:10.1002/qj.2748.

Schraff, C., M. Tomassini, and K. Stephan, 2008: Developments at dwd: Integrated water vapour (iwv) from ground-based gps. COSMO newsletter 9, Deutscher Wetterdienst, Deutscher Wetterdeinst, Frankfurter Strasse 135, 63067 Offenbach, Germany.

Schär, C., and Coauthors, 2016: Percentile indices for assessing changes in heavy precipitation events. *Climatic Change*, **137 (1-2)**, 201–216, doi:10.1007/s10584-016-1669-2.

Scoccimarro, E., S. Gualdi, A. Bellucci, M. Zampieri, and A. Navarra, 2013: Heavy precipitation events in a warmer climate: Results from CMIP5 models. *Journal of Climate*, **26 (20)**, 7902–7911, doi:10.1175/jcli-d-12-00850.1.

Seneviratne, S. I., T. Corti, E. L. Davin, M. Hirschi, E. B. Jaeger, I. Lehner, B. Orlowsky, and A. J. Teuling, 2010: Investigating soil moisture–climate interactions in a changing climate: A review. *Earth-Science Reviews*, **99 (3-4)**, 125–161, doi:10.1016/j.earscirev.2010.02.004.

Sherwood, S. C., R. Roca, T. M. Weckwerth, and N. G. Andronova, 2010: Tropospheric water vapor, convection, and climate. *Reviews of Geophysics*, **48 (2)**, doi:10.1029/2009rg000301.

Soderholm, J. S., H. A. McGowan, H. Richter, K. Walsh, T. Wedd, and T. M. Weckwerth, 2017: Diurnal preconditioning of subtropical coastal convective storm environments. *Monthly Weather Review*, **145 (9)**, 3839–3859, doi:10.1175/mwr-d-16-0330.1.

Steinke, S., 2017: Variability of integrated water vapour: An assessment on various scales with observations and model simulations over germany. Phd thesis, Universität zu Köln.

Steinke, S., S. Eikenberg, U. Löhnert, G. Dick, D. Klocke, P. D. Girolamo, and S. Crewell, 2014: Assessment of small-scale integrated water vapour variability during HOPE. *Atmospheric Chemistry and Physics Discussions*, **14 (16)**, 22 837–22 879, doi:10.5194/acpd-14-22837-2014.

Stevens, B., 2005: ATMOSPHERIC MOIST CONVECTION. *Annual Review of Earth and Planetary Sciences*, **33 (1)**, 605–643, doi:10.1146/annurev.earth.33.092203.122658.

Stull, R. B., Ed., 1988: *An Introduction to Boundary Layer Meteorology*. Springer Netherlands, doi:10.1007/978-94-009-3027-8.

Takayabu, Y. N., S. Shige, W.-K. Tao, and N. Hirota, 2010: Shallow and deep latent heating modes over tropical oceans observed with TRMM PR spectral latent heating data. *Journal of Climate*, **23 (8)**, 2030–2046, doi:10.1175/2009jcli3110.1.

Tiedtke, M., 1989: A comprehensive mass flux scheme for cumulus parameterization in large-scale models. *Monthly Weather Review*, **117 (8)**, 1779–1800, doi:10.1175/1520-0493(1989)117<1779:acmfsf>2.0.co;2.

Toreti, A., E. Xoplaki, D. Maraun, F. G. Kuglitsch, H. Wanner, and J. Luterbacher, 2010: Characterisation of extreme winter precipitation in mediterranean coastal sites and associated anomalous atmospheric circulation patterns. *Natural Hazards and Earth System Science*, **10 (5)**, 1037–1050, doi: 10.5194/nhess-10-1037-2010.

Tramblay, Y., and S. Somot, 2018: Future evolution of extreme precipitation in the mediterranean. *Climatic Change*, doi:10.1007/s10584-018-2300-5.

Vautard, R., and Coauthors, 2014: The european climate under a 2 °c global warming. *Environmental Research Letters*, **9 (3)**, 034 006, doi:10.1088/1748-9326/9/3/034006.

Verrelle, A., D. Ricard, and C. Lac, 2014: Sensitivity of high-resolution idealized simulations of thunderstorms to horizontal resolution and turbulence parametrization. *Quarterly Journal of the Royal Meteorological Society*, **141 (687)**, 433–448, doi:10.1002/qj.2363.

Weckwerth, T. M., 2000: The effect of small-scale moisture variability on thunderstorm initiation. *Monthly Weather Review*, **128 (12)**, 4017–4030, doi: 10.1175/1520-0493(2000)129<4017:teossm>2.0.co;2.

Weckwerth, T. M., J. W. Wilson, and R. M. Wakimoto, 1996: Thermodynamic variability within the convective boundary layer due to horizontal convective rolls. *Monthly Weather Review*, **124 (5)**, 769–784, doi:10.1175/1520-0493(1996)124<0769:tvwtcb>2.0.co;2.

Wernli, H., M. Paulat, M. Hagen, and C. Frei, 2008: SAL—a novel quality measure for the verification of quantitative precipitation forecasts. *Monthly Weather Review*, **136 (11)**, 4470–4487, doi:10.1175/2008mwr2415.1.

Wielicki, B. A., B. R. Barkstrom, E. F. Harrison, R. B. Lee, G. L. Smith, and J. E. Cooper, 1996: Clouds and the earth's radiant energy system (ceres): An earth observing system experiment. *Bulletin of the American Meteorological Society*, **77 (5)**, 853–868, doi:10.1175/1520-0477(1996)077<0853:CATERE>2.0.CO;2.

Winschall, A., H. Sodemann, S. Pfahl, and H. Wernli, 2014: How important is intensified evaporation for mediterranean precipitation extremes? *Journal of Geophysical Research: Atmospheres*, **119 (9)**, 5240–5256, doi:10.1002/2013jd021175.

WMO, 2012: Anticipated advances in numerical weather prediction (nwp), and the growing technology gap in weather forecasting. Technical report, World Meteorological Organization.

Zhang, G., K. H. Cook, and E. K. Vizy, 2016: The diurnal cycle of warm season rainfall over west africa. part II: Convection-permitting simulations. *Journal of Climate*, **29 (23)**, 8439–8454, doi:10.1175/jcli-d-15-0875.1.

Zhang, M., and D.-L. Zhang, 2012: Subkilometer simulation of a torrential-rain-producing mesoscale convective system in east china. part i: Model verification and convective organization. *Monthly Weather Review*, **140 (1)**, 184–201, doi:10.1175/mwr-d-11-00029.1.

Zhu, Y., and R. E. Newell, 1998: A proposed algorithm for moisture fluxes from atmospheric rivers. *Monthly Weather Review*, **126 (3)**, 725–735, doi:10.1175/1520-0493(1998)126<0725:apafmf>2.0.co;2.

Acknowledgement

I would like to thank first, Prof. Dr. Kottmeier for guiding this Ph.D Thesis and for his enlightening comments and advice. I would also like to thank Prof. Dr. Knippertz for co-examining this thesis and supervising my presentations.
I am wholeheartedly thankful to my supervisor Dr. Samiro Khodayar for the many fruitful discussions and guidance. Her clear vision and intelligence were crucial in bringing this thesis to meaningful results, never losing sight of what were the relevant aspects that could bring an impact to other scientists and future investigations. I could not put into words how grateful I am to have been given the chance to enjoy these years of work at the Institute for Meteorology and Climate Research (IMK) that have brought me so much professional and personal development. Mi mas sincero agradecimiento.
A special mention goes to Sebastian Helgert. The developed friendship during the extensive working hours, scientific sharing and personal support are, together with this thesis the most valuable rewards of this period. His calm intelligence and faith in the best, have been an inspiration to advance week after week.
Funding of this thesis and the publications in journals and international scientific conferences has been made possible thanks to the support program Junior Investigator group Global Change 4+1 (*Nachwuchsgruppen Globaler Wandel 4 + 1*) of the Federal Ministry of Education and Research (Bundesministerium für Bildung und Forschung; BMBF).
This thesis would not have been possible without the support of many colleagues that were always selflessly open to give a hand. Be it with scientific discussion and corrections or by sharing their very useful bits of code. I would

like to thank Dr. Hans-Jürgen Panitz for his support in running the COSMO in Climate code, and the modelleing experts at IMK, Bianca Adler and Christian Barthlott. Many thanks as well to Beata, Philip, Jutta, Karmen, Manuela, Yaoyao, Simon, Schweta, Lukas, Andrew, Rike, Ali, Dominick, Julian, Simone, Leo, Tobias, Norbert, Uli, Jan, Gabi, Nicole, George, Anubhav and Martin.

Very useful was the help offered by Johannes Hoerner and Alexander Reinbold as scientific support of the computing tasks in our working group. It was a very nice experience working together with M.Sc. students at the university.

Outside IMK, I have had the chance to share my research with renowned scientists and learn from their experience. I would like to thank the people at the German Weather Service (Deutsche Wetterdienst; DWD) for their time and answers, in particular to Dr. Christoph Schraff for his help in understanding and configurating the Nudging scheme, Dr. Ulrich Schättler for his help with the COSMO model, and the people at the Data Assimilation department, Dr. Elisabeth Bauernschubert and Prof. Dr. Roland Potthast. Within the HyMeX community, I would like to acknowledge Dr. Olivier Bock for providing the GPS-derived ZTD data set and Stefano Federico and Silvio Davolio for the very fruitful scientific exchange at the HyMeX workshops.

In a more personal sphere, I would like to thank my parents Marisa Alvarez and Antonio Caldas for their advice and love. You are my inspiration and my soul. When you talk, the world comes to an order. To my beloved sisters, Maria and Beatriz. Your intelligence leaves me in awe. I admire you every day.

I would like to thank as well my second family for all their support and love during these years. Lila, Nacho, Lena, Sebastian and Jasi, I feel very lucky to have come across your paths. Also I would like to mention all the friends that made of Karlsruhe the city of light and opportunities that it is. My best thoughts to William, Marc, Ester Diaz, Luis Manuel, Alejandro, Julia, Thorsten, Mario Kiermeier, Karin, Luis Luengo, Adrian Juan, Alvaro, Jose Tijero, Sebastien, Chiara, Grecia, Edgar Berrospe, Pancho, Emanuelle, Oli, Daniel, Engin and the JRC footballers, Matthias, Darko, Flo, Jenni, Matthieu, Santiago Ochoa, Mario Rodrigues and Lisa.

The warmth of the southwesterly flows could also be felt in Karlsruhe. Even though virtually I was also lucky to count on Susana, Adrian Otegui, Julio, Mayor, Correa, Villota, Barbara, Clara, Jorge, Oscar and Ester Aranzana, Oreja, Guti, Bengo and Pablito. And from my ever-remebemerd Erasmus friends Marta, Hector, Olmo, Baptiste and Florian.

Wissenschaftliche Berichte des Instituts für Meteorologie und Klimaforschung des Karlsruher Instituts für Technologie (0179-5619)

Bisher erschienen:

Nr. 1: *Fiedler, F. / Prenosil, T.*
Das MESOKLIP-Experiment. (Mesoskaliges Klimaprogramm im Oberrheintal). August 1980

Nr. 2: *Tangermann-Dlugi, G.*
Numerische Simulationen atmosphärischer Grenzschichtströmungen über langgestreckten mesoskaligen Hügelketten bei neutraler thermischer Schichtung. August 1982

Nr. 3: *Witte, N.*
Ein numerisches Modell des Wärmehaushalts fließender Gewässer unter Berücksichtigung thermischer Eingriffe. Dezember 1982

Nr. 4: *Fiedler, F. / Höschele, K. (Hrsg.)*
Prof. Dr. Max Diem zum 70. Geburtstag. Februar 1983 (vergriffen)

Nr. 5: *Adrian, G.*
Ein Initialisierungsverfahren für numerische mesoskalige Strömungs modelle. Juli 1985

Nr. 6: *Dorwarth, G.*
Numerische Berechnung des Druckiderstandes typischer Geländeformen. Januar 1986

Nr. 7: *Vogel, B.; Adrian, G. / Fiedler, F.*
MESOKLIP-Analysen der meteorologischen Beobachtungen von mesoskaligen Phänomenen im Oberrheingraben. November 1987

Nr. 8: *Hugelmann, C.-P.*
Differenzenverfahren zur Behandlung der Advektion. Februar 1988

Nr. 9: *Hafner, T.*
Experimentelle Untersuchung zum Druckwiderstand der Alpen. April 1988

Nr. 10: *Corsmeier, U.*
Analyse turbulenter Bewegungsvorgänge in der maritimen atmosphärischen Grenzschicht. Mai 1988

Nr. 11: *Walk, O. / Wieringa, J.(eds)*
Tsumeb Studies of the Tropical Boundary-Layer Climate. Juli 1988

Nr. 12: *Degrazia, G. A.*
Anwendung von Ähnlichkeitsverfahren auf die turbulente Diffusion in der konvektiven und stabilen Grenzschicht. Januar 1989

Nr. 13: *Schädler, G.*
Numerische Simulationen zur Wechselwirkung zwischen Landoberflächen und atmophärischer Grenzschicht. November 1990

Nr. 14: *Heldt, K.*
Untersuchungen zur Überströmung eines mikroskaligen Hindernisses in der Atmosphäre. Juli 1991

Nr. 15: *Vogel, H.*
Verteilungen reaktiver Luftbeimengungen im Lee einer Stadt – Numerische Untersuchungen der relevanten Prozesse. Juli 1991

Nr. 16: *Höschele, K.(ed.)*
Planning Applications of Urban and Building Climatology – Proceedings of the IFHP / CIB-Symposium Berlin, October 14-15, 1991. März 1992

Nr. 17: *Frank, H. P.*
Grenzschichtstruktur in Fronten. März 1992

Nr. 18: *Müller, A.*
Parallelisierung numerischer Verfahren zur Beschreibung von Ausbreitungs- und chemischen Umwandlungsprozessen in der atmosphärischen Grenzschicht. Februar 1996

Nr. 19: *Lenz, C.-J.*
Energieumsetzungen an der Erdoberfläche in gegliedertem Gelände. Juni 1996

Nr. 20: *Schwartz, A.*
Numerische Simulationen zur Massenbilanz chemisch reaktiver Substanzen im mesoskaligen Bereich. November 1996

Nr. 21: *Beheng, K. D.*
Professor Dr. Franz Fiedler zum 60. Geburtstag. Januar 1998

Nr. 22: *Niemann, V.*
Numerische Simulation turbulenter Scherströmungen mit einem Kaskadenmodell. April 1998

Nr. 23: *Koßmann, M.*
Einfluß orographisch induzierter Transportprozesse auf die Struktur der atmosphärischen Grenzschicht und die Verteilung von Spurengasen. April 1998

Nr. 24: *Baldauf, M.*
Die effektive Rauhigkeit über komplexem Gelände – Ein Störungstheoretischer Ansatz. Juni 1998

Nr. 25: *Noppel, H.*
Untersuchung des vertikalen Wärmetransports durch die Hangwindzirkulation auf regionaler Skala. Dezember 1999

Nr. 26: *Kuntze, K.*
Vertikaler Austausch und chemische Umwandlung von Spurenstoffen über topographisch gegliedertem Gelände. Oktober 2001

Nr. 27: *Wilms-Grabe, W.*
Vierdimensionale Datenassimilation als Methode zur Kopplung zweier verschiedenskaliger meteorologischer Modellsysteme. Oktober 2001

Nr. 28: *Grabe, F.*
Simulation der Wechselwirkung zwischen Atmosphäre, Vegetation und Erdoberfläche bei Verwendung unterschiedlicher Parametrisierungsansätze. Januar 2002

Nr. 29: *Riemer, N.*
Numerische Simulationen zur Wirkung des Aerosols auf die troposphärische Chemie und die Sichtweite. Mai 2002

Nr. 30: *Braun, F. J.*
Mesoskalige Modellierung der Bodenhydrologie.
Dezember 2002

Nr. 31: *Kunz, M.*
Simulation von Starkniederschlägen mit langer Andauer über Mittelgebirgen. März 2003

Nr. 32: *Bäumer, D.*
Transport und chemische Umwandlung von Luftschadstoffen im Nahbereich von Autobahnen – numerische Simulationen.
Juni 2003

Nr. 33: *Barthlott, C.*
Kohärente Wirbelstrukturen in der atmosphärischen Grenzschicht. Juni 2003

Nr. 34: *Wieser, A.*
Messung turbulenter Spurengasflüsse vom Flugzeug aus.
Januar 2005

Nr. 35: *Blahak, U.*
Analyse des Extinktionseffektes bei Niederschlagsmessungen mit einem C-Band Radar anhand von Simulation und Messung.
Februar 2005

Nr. 36: *Bertram, I.*
Bestimmung der Wasser- und Eismasse hochreichender konvektiver Wolken anhand von Radardaten, Modellergebnissen und konzeptioneller Betrachtungen. Mai 2005

Nr. 37: *Schmoeckel, J.*
Orographischer Einfluss auf die Strömung abgeleitet aus Sturmschäden im Schwarzwald während des Orkans „Lothar". Mai 2006

Nr. 38: *Schmitt, C.*
Interannual Variability in Antarctic Sea Ice Motion: Interannuelle Variabilität antarktischer Meereis-Drift.
Mai 2006

Nr. 39: *Hasel, M.*
Strukturmerkmale und Modelldarstellung der Konvektion über Mittelgebirgen. Juli 2006

Ab Band 40 erscheinen die Wissenschaftlichen Berichte des Instituts für Meteorologie und Klimaforschung bei KIT Scientific Publishing (ISSN 0179-5619). Die Bände sind unter www.ksp.kit.edu als PDF frei verfügbar oder als Druckausgabe bestellbar.

Nr. 40: *Lux, R.*
Modellsimulationen zur Strömungsverstärkung von orographischen Grundstrukturen bei Sturmsituationen
ISBN 978-3-86644-140-8

Nr. 41: *Straub, W.*
Der Einfluss von Gebirgswellen auf die Initiierung und Entwicklung konvektiver Wolken
ISBN 978-3-86644-226-9

Nr. 42: *Meißner, C.*
High-resolution sensitivity studies with the regional climate model COSMO-CLM
ISBN 978-3-86644-228-3

Nr. 43: *Höpfner, M.*
Charakterisierung polarer stratosphärischer Wolken mittels hochauflösender Infrarotspektroskopie
ISBN 978-3-86644-294-8

Nr. 44: *Rings, J.*
Monitoring the water content evolution of dikes
ISBN 978-3-86644-321-1

Nr. 45: *Riemer, M.*
Außertropische Umwandlung tropischer Wirbelstürme: Einfluss auf das Strömungsmuster in den mittleren Breiten
ISBN 978-3-86644-766-0

Nr. 46: Nicht erschienen.

Nr. 47: Nicht erschienen.

Nr. 48: Nicht erschienen.

Nr. 49: *Peters, T.*
Ableitung einer Beziehung zwischen der Radarreflektivität, der Niederschlagsrate und weiteren aus Radardaten abgeleiteten Parametern unter Verwendung von Methoden der multivariaten Statistik
ISBN 978-3-86644-323-5

Nr. 50: *Khodayar Pardo, S.*
High-resolution analysis of the initiation of deep convection forced by boundary-layer processes
ISBN 978-3-86644-770-7

Nr. 51: *Träumner, K.*
Einmischprozesse am Oberrand der konvektiven atmosphärischen Grenzschicht
ISBN 978-3-86644-771-4

Nr. 52: Nicht erschienen.

Nr. 53: *Lundgren, K.*
Direct Radiative Effects of Sea Salt on the Regional Scale
ISBN 978-3-86644-773-8

Nr. 54: *Sasse, R.*
Analyse des regionalen atmosphärischen Wasserhaushalts unter Verwendung von COSMO-Simulationen und GPS-Beobachtungen
ISBN 978-3-86644-774-5

Nr. 55: *Grenzhäuser, J.*
Entwicklung neuartiger Mess- und Auswertungsstrategien für ein scannendes Wolkenradar und deren Anwendungsbereiche
ISBN 978-3-86644-775-2

Nr. 56: *Grams, C.*
Quantification of the downstream impact of extratropical transition for Typhoon Jangmi and other case studies
ISBN 978-3-86644-776-9

Nr. 57: *Keller, J.*
Diagnosing the Downstream Impact of Extratropical Transition Using Multimodel Operational Ensemble Prediction Systems
ISBN 978-3-86644-984-8

Nr. 58: *Mohr, S.*
Änderung des Gewitter- und Hagelpotentials im Klimawandel
ISBN 978-3-86644-994-7

Nr. 59: *Puskeiler, M.*
Radarbasierte Analyse der Hagelgefährdung in Deutschland
ISBN 978-3-7315-0028-5

Nr. 60: *Zeng, Y.*
Efficient Radar Forward Operator for Operational Data Assimilation within the COSMO-model
ISBN 978-3-7315-0128-2

Nr. 61: Nicht erschienen.

Nr. 62: *Jerger, D.*
Radar Forward Operator for Verification of Cloud Resolving Simulations within the COSMO Model
ISBN 978-3-7315-0172-5

Nr. 63: *Maurer, V.*
Vorhersagbarkeit konvektiver Niederschläge : Hochauflösende Ensemblesimulationen für Westafrika
ISBN 978-3-7315-0189-3

Nr. 64: *Stawiarski, C.*
Optimizing Dual-Doppler Lidar Measurements of Surface Layer Coherent Structures with Large-Eddy Simulations
ISBN 978-3-7315-0197-8

Nr. 65: *Mahlke, H.*
Mechanismen der Auslösung hochreichender Konvektion im südwestdeutschen Mittelgebirgsraum
ISBN 978-3-7315-0203-6

Nr. 66: *Fosser, G.*
Precipitation statistics from regional climate model at resolutions relevant for soil erosion
ISBN 978-3-7315-0227-2

Nr. 67: *Adler, B.*
Boundary-Layer Processes Producing Mesoscale Water-Vapour Variability over a Mountainous Island
ISBN 978-3-7315-0247-0

Nr. 68: *Kraut, I.*
Separating the Aerosol Effect in Case of a „Medicane"
ISBN 978-3-7315-0405-4

Nr. 69: *Breil, M.*
Einfluss der Boden-Vegetation-Atmosphären Wechselwirkungen auf die dekadische Vorhersagbarkeit des Westafrikanischen Monsuns
ISBN 978-3-7315-0420-7

Nr. 70: *Lott, F. F.*
Wind Systems in the Dead Sea and Footprints in Seismic Records
ISBN 978-3-7315-0596-9

Nr. 71: *Rieger, D.*
Der Einfluss von natürlichem Aerosol auf Wolken über Mitteleuropa
ISBN 978-3-7315-0672-0

Nr. 72: *Loewe, K.*
Arctic mixed-phase clouds. Macro- and microphysical insights with a numerical model
ISBN 978-3-7315-0686-7

Nr. 73: *Piper, D. A.*
Untersuchung der Gewitteraktivität und der relevanten großräumigen Steuerungsmechanismen über Mittel- und Westeuropa
ISBN 978-3-7315-0701-7

Nr. 74: *Metzger, J.*
Wind Systems and Energy Balance in the Dead Sea Valley
ISBN 978-3-7315-0699-7

Nr. 75: *Deetz, K.*
Assessing the Aerosol Impact on Southern West African Clouds and Atmospheric Dynamics
ISBN 978-3-7315-0744-4

Nr. 76: *Ehmele, F. T.*
Stochastische Simulation großflächiger, hochwasserrelevanter Niederschlagsereignisse.
ISBN 978-3-7315-0761-1

Nr. 77: *Hackenbruch, J.*
Anpassungsrelevante Klimaänderungen für städtische Baustrukturen und Wohnquartiere
ISBN 978-3-7315-0771-0

Nr. 78: *Schmidberger, M.*
Hagelgefährdung und Hagelrisiko in Deutschland basierend auf einer Kombination von Radardaten und Versicherungsdaten.
ISBN 978-3-7315-0846-5

Nr. 79: *Gruber, S.*
Contrails and Climate Engineering - Process Studies on Natural and Artificial High-Level Clouds and Their Impact on the Radiative Fluxes.
ISBN 978-3-7315-0896-0

Nr. 80: *Walter, C.*
Simulationen der Ausbreitung von Vulkanasche unter expliziter Berücksichtigung der optischen Eigenschaften der Aschepartikel.
ISBN 978-3-7315-0939-4

Nr. 81: *Caldas-Alvarez, A.*
Atmospheric Moisture Effects on Deep Convection in the Western Mediterranean.
ISBN 978-3-7315-0947-9